Codes, Puzzles, and Conspiracy

Codes, Puzzles, and Conspiracy

Dennis Shasha

W. H. Freeman and Company
NEW YORK

Chapter opening illustrations by Warren Linn.

Library of Congress Cataloging-in-Publication Data

Shasha, Dennis Elliott.

Codes, puzzles, and conspiracy / Dennis Shasha.

p. cm.

ISBN 0-7167-2314-X (hbk.)

ISBN 0-7167-2275-5 (pbk.)

1. Mathematical recreations. I. Title.

QA95.S469 1992

793.7′4—dc20 91-24223

CIP

Printed in the United States of America

1 2 3 4 5 6 7 8 9 0 VB 9 9 8 7 6 5 4 3 2 1

To Jeff, Ari, Cloe, Nick, and Isaac—
among the next generation of omniheurists

Contents

The following symbols indicate the level of difficulty of the puzzles: ● = easy; ■ = intermediate; ◆ = expert; ◆◆ = very expert.

PREFACE xi

ACKNOWLEDGMENTS xii

THE CODED LETTER 1

1. The Letter ● 3

ARMADA AT MOHER 9

2. Running Bulls ■ 11
3. Humpty Dumpty at the Irish Cliffs ■ 17
4. El Casino ■ 19
5. Greed and Getaways ■ 22
6. Museum Tour ■ 26

THE GENIUS OF GEORGETOWN 29

7. Musical Messages ◆ 31
8. Elves Flip ◆ 36
9. A Problem of Protocol ■ 40
10. Tropical Antarctica ◆ 45
11. 100-Day Rockets ◆ 48
12. Three-Finger Shooting ■ 51

JUSTIFYING THE MEANS 55
13. Sand Magic ■ 57
14. Signals and Echoes ● 60
15. The Territory Game ■ 64
16. Drugs and Interdiction ● 67

JUNGLE KILLERS 71
17. Amazon Exchange ■ 73
18. Mutual Admiration ■ 77

DOUBLE ESCAPE 81
19. The Octopelago Problem ● 83
20. MicroAir ■ 86
21. Personals ● 90
22. Shark Labyrinth ■ 92
23. A Question of Inheritance ◆◆ 97
24. The Toxicologist's Puzzle ■ 100

BATTLE FOR A CONTINENT 107
25. How to Steal a Submarine ■ 109
26. Missile Roulette ■ 114
27. Finding the Target ■ 118

UNEASY PEACE 123
28. Epidemiologists ● 125
29. MarsRail ◆◆ 128
30. The Prince's Problem ■ 130
31. Oil and Water ■ 133
32. Pomp But No Power? ■ 136
33. Power Grab ■ 138

A SECRET SOCIETY 143
34. Odd Voters ◆ 145
35. Polling the Oddists ◆ 149

36. The Hokkaido Post Office Problem ■ 152
37. Joining the Oddists ■ 156
38. Oddist Summer Training ◆ 159

CONCENTRIC CONSPIRACIES 163
39. Plea from a Fugitive ■ 165
40. The Prosecution Makes Its Case ■ 166
41. Friends in High Places ● 173
42. Nightly News ■ 177
43. Television Treason ◆ 180
44. Peirce's Beanbag ◆ 182
45. Three Notes ■ 190

SOLUTIONS TO THE PUZZLES 193

Preface

Imagination is more
important than knowledge.
Albert Einstein

Great doctors, physicists, sculptors, construction workers, and computer scientists tell me the same story: the joy in their professions comes from solving a problem that a moment before seemed unsolvable. Sometimes they go on to say that they are successful because they are good at solving puzzles. But career rewards, they assert, are secondary. The thrill belongs to the struggle—and triumph.

Like its predecessor *The Puzzling Adventures of Dr. Ecco,* this book is dedicated to people who love to think and enjoy thrillers. Ecco, Evangeline, and Scarlet use mathematics and logic to battle a hidden conspiracy, and you are invited to enter the fray. The story revolves around original mathematical puzzles that you are as well equipped to solve as the protagonists.

The puzzles themselves take their inspiration from central ideas in modern mathematics and computer science (graph theory, distributed computing, design theory, cryptography, and so on), but you do not need a background in any of these subjects. Some of the best puzzlers I know are artists (my wife, Karen), not yet teenagers (my neighbor, Isaac), or self-taught inventors (my friend, Steve). You need insight and imagination to solve puzzles, not necessarily knowledge. If you do solve the 45 puzzles in this book, you will learn a lot about all of the fields mentioned. I don't mean facts or terms, but rather ideas and ways of imagining. That is, you will learn what is important.

Acknowledgments

My wife Karen is the best intuitive puzzler I know, though she insists on remaining an artist. Without her contributions—spiritual, aesthetic, and mathematical—this book would not have been possible.

Amanda Brauman, Brad Barber, and Fred Galvin read the manuscript in various stages of its development and spent many hours making constructive and extremely helpful comments.

My extended family Hanina, Alfred, Victor, Angela, Carol, Robert, Edward, Michael, and Carrie helped with their many suggestions and moral support. Angela and Edward deserve special thanks for reading the entire manuscript.

Colleagues at universities, AT&T Bell Labs, and friends have, sometimes unwittingly, contributed to this book. I would particularly like to thank: Magdy Abadir, Don Beaver, Marc Donner, Bill Griffeth, Cornelius Groenewoud, Stuart Haber, Lynne Hale, Keith Heard, David Heilbroner, Eric Jordan, Henk Klunder, Hosam Mahmoud, Bill Mason, Colm O'Dunlaing, Ricky Pollack, Mortimer Propp, Eric Weeks, and Gunning Point.

My editor Jerry Lyons, project editor Christine Hastings, copyeditor Nancy Brooks, designer Alison Lew, illustration coordinator Bill Page, and the many other talented people at Freeman made this book a pleasure to produce.

Several of these puzzles have appeared in various guises in the recreational programming newsletter *Algorithm* (Algorithm Publications, 362 Wortley Road, London, Ontario N6C 3S2 Canada). Many thanks to the entire editorial staff, particularly A. K. Dewdney, and to the intrepid readers who attempted Ecco's puzzles in that form.

Finally, I would like to thank J. E. for his (grudging) permission to record his adventures.

The Second Omniheurist's Contest

Scattered throughout this book are five contest puzzles and one optional one. A convincing answer to the optional puzzle will mitigate an incorrect answer to the five mandatory ones. In the back of the book, you will find complete directions for submitting your solutions as entries in the contest. Good luck.

The Coded Letter

1. The Letter

Whether the jug hits the stone
or the stone hits the jug,
it is bad for the jug.
Folk saying

More than two years had passed since I had seen or heard from Ecco. I stayed in touch with Evangeline, of course, and each weekend she came up to New York from Princeton. Together we'd pour over newspapers, magazines, and mail searching for a significant clue. So far, nothing.

Many people would have called us eggheads: Dr. Jacob Ecco, mathematical prodigy and the world's foremost omniheurist; Dr. Evangeline Goode, Princeton philosopher with a special interest in the logic of learning; and I, Professor Justin Scarlet of the Mathematics Institute here in New York, my field topology of structures.

By definition, an omniheurist is one who can solve all problems. That sounds like the claim of an advice columnist, but recall the Latin and Greek roots of the word: *omnia*, all things; ***heuriskein***, to find out. As developed by our friend, the science of omniheuristics applies a mathematical approach to the solving of problems. Omniheuristics is more a method, a way of seeing things, than a specific field of study. Specialists may supply critical data, but the generalist omniheurist, through insight and reason alone, finds the solution. The very word was Ecco's invention, and he was the first practitioner of the technique. He had employed it to help repair malfunctioning spacecraft, to catch spies, to design buildings. He had even used it in meeting challenges thrown down by the notorious and illusive Benjamin Baskerhound, a rogue genius and erstwhile Princeton professor whose warped love of puzzles had led him to crime. Futurists have acclaimed omniheurism as a science for the new millennium, pointing out that now that terrestrial frontiers have nearly vanished, the problems we face as human beings arise from the objects, systems, and relationships we ourselves create.

Ecco, Evangeline, and I had had exciting intellectual adventures together, which I had the privilege of chronicling for the public in *The*

Puzzling Adventures of Dr. Ecco, and some good times windsurfing. Always, Jacob Ecco was our rigorous mentor and good friend. He was a quiet hero, a man who defied generals and blackmailing criminals alike if what they said was nonsense, untrue, or wrong. We missed his fine mind, his indifference to fame, his love of chess, his passion for cookies.

From the beginning of our search there had been offers of help, well meant, no doubt, but useless. The Director, with whom we had worked in some matters of national security, had told us he would put agents at our disposal if ever we had a promising lead. Personally, his sympathy meant little to Evangeline and me: the Director had a manner that was always arrogant and offensively mysterious, and neither of us liked him.

When Ecco first disappeared, the press was full of stories of the most speçulative kind. One actually proposed that Ecco had been sent up in a secret "Encounters Satellite" to communicate with aliens; others traced sightings of him in all corners of the globe, from Tibet to Tierra del Fuego. Usually, a few telephone inquiries were enough to convince us that someone had made a mistake or was playing a prank. We made several trips, as far as Delhi and Buenos Aires, but in every case the "evidence" seemed to evaporate when we arrived. After a time, the press lost interest, but the theories and tips of well-intentioned individuals kept us busy. The result—or lack of it—was always the same.

On this Sunday, the only relevant news was a piece in the *New York Times Sunday Magazine* by Cloe Anne Bennet about the fabulously successful Omniheurism Inc., a consulting firm that had risen to prominence shortly after Ecco's disappearance.

"Since a client often puts the omniheurist's advice immediately to the test in matters involving great sums of money, hundreds of people, and huge projects, failure can be catastrophic," she wrote. "Success is all or nothing." She went on:

> Only a handful of companies and individuals have survived this daunting challenge. Of these, Omniheurism Inc. has come to be the first, and certainly the priciest, choice of civilian and military government officials as well as of wealthy corporations and individuals.
>
> Imagine, if you will, a stately mansion in the elegant Georgetown section of Washington, D.C. A butler escorts the visitor to a large domed

room, originally a ballroom. The room is unfurnished except for a single oak desk and a few chairs. Seated at the desk is Dr. Phillip Andrew Smartee, a graduate of Eton and Christ Church, Oxford.

A tall, handsome man, impressively tailored, Dr. Smartee begins all interviews by asking his clients to state their problem. Dr. Smartee speaks so little at this stage that the interaction resembles a psychoanalytic session. On the rare occasions when he has a question, he raises his index finger. When the client stops talking, Dr. Smartee states his question quietly, almost in a whisper. After the problem has been presented to his satisfaction, he raises his finger again and walks through a door in the back of the ballroom. The door leads to what he calls his Idea Chamber.

Sometimes he emerges to ask a few questions, listening to the answers as if deep in thought. Then he returns to the Idea Chamber. When he reemerges, he either provides a solution on the spot or asks the client to return in a day or so for the solution.

On my first visit, Dr. Smartee described to an important Hollywood producer how to arrange film shots so as to minimize the expense of keeping famous—and highly paid—actors waiting. On my second visit, he suggested an elevator design to Takomoto, the construction company whose plans for a 200+ story office building had just been announced. Dr. Smartee's design allows elevator cars to change tracks the way trains do . . .

Evangeline slowly put the paper down. "Professor Scarlet, will we ever see Jacob again?" Her black almond-shaped eyes filled with tears. Her face showed the heritage of both her parents: she had the straight narrow nose of her American missionary father and the fine-textured, delicate skin of her Chinese mother. Now, in her pain over the loss of Ecco, her Chinese ancestry seemed the stronger, her face the mask of a monk in deep meditation.

"Of course we will," I said. "Ecco is alive and well." I knew my voice lacked conviction.

I looked back at the article. Bennet did not give full details about the various problems presented, but Smartee's solutions bore an uncanny resemblance to Ecco's work, though Smartee's fees were far higher. Bennet had tried to ask this newly famous omniheurist about his background, but Smartee would answer no personal questions. The reporter had taken her investigation to the Pentagon, figuring that the military must have examined Smartee's background before entrusting him with top-secret material, but much of their information about him was classified. Still, Bennet had learned a few interesting things. Apparently, "Smartee" was a pseudonym, although she quoted a master at Eton as saying, "If this is the boy I think, we must have been utterly wrong about him. Though well-born and vain, he never struck me as intelligent."

As we discussed the article, Evangeline and I tried to encourage each other. But the last months had been difficult, and Evangeline's question showed we were both beginning to think what we had never admitted even to ourselves—that we would never see Ecco again.

"How stupid of me!" I exclaimed in an effort to be positive and banish such thoughts. "I forgot to check the mail yesterday." For two years I had instilled every trip to my mailbox with hope for a good lead. But two years of trips to the mailbox had resulted in nothing but disappointment. Why should today be different?

Evangeline went downstairs with me. The box was full of junk mail, but there was one air mail envelope, tattered at the sides. The ink had run, indicating it had been wet. However, its contents were perfectly legible, and we opened it to recognize the unmistakable hand of Jacob Ecco:

My Dear Evangeline and Scarlet,

Fazr wxxkiqxn xa qocu zq if xrwoh mwek dkkb johotkbx, xmaztm wccarjobt xa if cwqxar, hwztmwdhf obcaiqkxkbx. Qhkwnk jab'x xwuk appkbnk, par mk on wb obxahkrwdhk nbad.

Dkcwznk mk on znobt ik xa nahek iwbf joppoczhx qzyyhkn w jwf, mk toekn ik qroewcf wbj xoik xa xmobu. Xmk rknzhx on xmwx O cwb lroxk xa faz wx if hkonzrk. O mwek if jazdxn xmwx xmon hkxxkr lohh rkwcm faz, dzx O mwek w qhwb xa hkwek ox ob w trackr'n nxacu raai.

Am, fkn, if cwqxar on Dkbgwiob Dwnukrmazbj, xmk iwxm qrajotf lma dkcwik w qmohanaqmf qrapknnar wx Qrobckxab, bal xzrbkj uojbwqqkr. Mk iwf kbxkr axmkr cwrkkrn ekrf naab. Loxm xmk habt-lobjkj obxrajzcxoab azx ap xmk lwf, faz lohh pobj ox kwnf xa jkcajk xmk rknx ob nqoxk ap xmk pwcx xmwx xmk cajk on raxwxobt df abk hkxxkr rotmx bal.

Exovlsnback xck npo oyxqq xsl dbcypcaxiig bc ynl jbfl. Nl dbcysbio npo xqqxpso eg dliiaixs ylilrnbcl, aopcu rspfxyl oxyliipyl dnxcclio. Nl qpcko slxkg mlidbjlo pc ixcko mnbol rlbril xkjpsl slelio. Ml xsl cbm pc Asauaxg, x ixck qpiilk mpyn lwpilk Exotalo mnb nxk qbauny xuxpcoy Qsxcdb pc Orxpc. Elqbsl ynpo, ml mlsl pc Nbrp dbacysg mnlsl ynl ysxkpypbcxi oaorpdpbc ybmxsko bayopklso mxo kporliilk mnlc Exovlsnback (xck P) nlirlk yb ibdxyl x upxcy xtapqls. Ynl dbkl po sbyxypcu xuxpc. Nm pzyh qd y sjyem rct y lyh, y nmmw, y kcdzo. Zomd nm ytm crr, bpbyjjh qd Fypwmtocbdl'p JmytImzp.

Fypwmtocbdl qp pc ecdrqlmdz cr oqp yfqjqzh zc mgylm lmzmezqcd zoyz om yjjcnp km zc sjyh zom zcbtqpz--yeecksydqml fh kh "fclhvbytlp," cr ecbtpm. Zom eclm qp tczyzqdv.

Kddx sdu ln re Cucwczi. R orkk aui ad srwcun dca dcu enya mnqarezarde zem orkk knzhn odum orap z fpnnusck zem caanuki ezrhn pduqn-gunnmnu de z mrua udzm enzu z hrkkzwn fzkknm Tceaz Gzkknez. Zqx sdu apn truzan'q pdcqn zem idc ode'a gn szu zozi. Opne idc srem apn pduqn-gunnmnu, zqx prl opnapnu pn rq uzfrew zei kzln pduqnq aprq inzu. Fdmn fpzewnq zwzre. S psll feb ho ahlo be dro bqsr geno axasf.

Bqo ofgensfx et bqo morraxo S loaio fozb bsmo psll vebabo efgo oagq bsmo bqo morraxo gefbasfr bqo lobbov . . . poll, jed psll tsxdvo sb edb. Bqo bsmo atbov bqab, bqo geno psll ho bqo ramo efo bqab bqsr lobbov hoxaf psbq, ozgoub bqab bqo morraxo psll ho rulsb sfbe xvedur et osxqb gqavagbovr. Bqoro xvedur psll ho pvsbbof sf vafnem evnov, hdb remo et bqom eiovlau.

Haryovqedfn psll ho wdsgylj apavo et afj makev uelsgo euovabsef, re uloaro ho rdhblo. Lob remoefo yfep pqovo jed avo xesfx sf garo jed nsrauuoav.

Jedvr,
Kageh

Armada at Moher

2. Running Bulls

Be good and you will be lonesome.
Mark Twain

Evangeline and I took several hours to decode Ecco's letter. We were confused by the rotations, but we understood our friend's motivation and were grateful for his hints about the coding of future messages. I for one wondered whether there would be any future messages.

Ecco wanted to be rescued, there was no doubt of that. But rescues are dangerous and Baskerhound was a formidable opponent. We decided to suppress our dislike of the Director and ask for his help. He had given us a phone number and told us that an answering machine would take our message. If ever we had a lead, we should say, "Jake is awake." We made the call. Within twenty-four hours, the Director came to see us, accompanied by security men. He had not changed much in the past two years. He still carried his six feet two well, and most people would consider him handsome, with his strong jaw and chiseled features. But I always found myself looking at his small, humorless eyes.

"I hope you have a good lead," he said after a short nod of his head that passed for a greeting. "We could use Ecco's help in tracking down some international anarchists. Smartee is no substitute for the real thing. What have you heard?"

We told him that we had received a letter from Ecco and were going to Uruguay. To our surprise, he didn't ask to see the letter. He told us simply that we would have immediate help if we phoned the same number and left the message, "For Jake's sake."

"And if we can't get to a phone?" Evangeline asked.

"My dear Dr. Goode," said the Director, "we have already thought of that." He held up a small metallic object the size and shape of a credit card. "This is a solar-powered transmitter that sends out a distress signal. Our agents call it a pip-card. It emits a low-power signal that our planes, with their sensitive receivers, will pick up provided they are within two hundred miles of you. If you find yourself in trouble, put the pip-card out during the day or under the light of a lamp. Understand?" He handed the card to Evangeline and left. The next evening we took a flight to Montevideo.

Punta Ballena is a ridge rising from the mouth of the Rio de la Plata. Approached from the sea, it looks like the profile of a whale (*ballena* is the Spanish word for "whale"). The spine of the ridge is the area's paved road. Dirt paths lead from the main road to private farms and ranches, some expensive and well kept, but most ramshackle.

We tracked down the horse-breeder with only minor difficulty, finding him in the stables of one of the largest estates in Punta Ballena. When we asked him in halting Spanish whether he raced his lame horses, he laughed and invited us in for "maté," pointing at his mug of thick tea. "*Es bueno,*" he assured us, "*muy fuerte.*"

The horse-breeder told us about the "crazy gringo." "He takes long rides in the arboretum, sometimes coming back with plant clippings. Whenever he is around, the arboretum dog gets sick and vomits. A week ago, the dog nearly died. Sometimes the gringo brings a younger red-haired man with him. The younger man is almost a prisoner. He is always surrounded by big men with small heads, *me comprende?*"

Evangeline took out a color snapshot of Ecco, his red hair very evident in the sunlight. "Si, es el rojo," said the horse-breeder.

"They went away," he said. "Last Saturday. *El hombre* with the red hair left you this." He handed Evangeline a letter. We thanked him, relieved that we didn't have to drink more maté, which tasted to me like hot wet grass. As we were saying our farewells, an exquisitely dressed gentleman, the picture of a Latin aristocrat, came riding up on a stallion. Our host raced to meet him as he dismounted.

"Pedro Alcatraz is my name, honored *señor y señora,*" the newcomer said with a bow to Evangeline. "I have just heard that you are friends of the poor red-haired *muchacho* whom that crazy Baskerhound is keeping prisoner. I have heard rumors that this red-haired man is the great Dr. Jacob Ecco and that you are his colleagues, Professor Justin Scarlet and Dr. Evangeline Goode. Is this true?"

We both felt that a denial would arouse more suspicion than it would allay, so we nodded hesitantly.

"Please do not worry. I hate that Baskerhound. He is always digging up the plants of my family's arboretum. You see, I am the great-grandson of the great corsair."

He smiled at our surprise. "What is the use of protesting my great-grandfather's innocence? In our family, small boys play 'hide the stolen treasure.' My great-grandfather made his money by turning off the light in the local lighthouse during storms and waiting for shipwrecks. He used the money to collect rare plants and trees. Somewhere in the arboretum is a chest of rare coins, too, but we have never found it. For a time, I suspected Baskerhound was looking for that chest, but he never digs deeply enough.

"Anyway, we have used our wealth for many humanitarian causes in the last few generations. Perhaps you will help me accomplish the goal of my own humble lifetime: to eliminate bullfighting in Uruguay."

Evangeline and I exchanged glances.

"Oh, I've tried everything," Señor Alcatraz said. "Petitions to lawmakers, newspaper advertisements, everything. But the people aren't with me. They need the entertainment. I have given up moralizing. My goal now is to interest the people in bull races instead of bullfighting. Will you help me?"

"We can try," I said uncertainly.

"I am sure you can help," said Señor Alcatraz. "Let me tell you a little history. Two years ago, I built four parallel running tracks on my large *estancia* near Montevideo. My idea was to release four bulls simultaneously at the start of each track. Each would run to the end of its track. I built a high stadium around the tracks, so the spectators could watch the bulls. Unfortunately, the people found it boring to watch four bulls on four independent tracks. So, I built cross-tracks that allow the bulls to go in any direction they like. The trouble is that the bulls have a terrible habit of fighting with one another. The people love this, *Dios Mio,* but it is very expensive and defeats the point of my efforts." He shrugged. "One of the fans suggested that I create races in which the bulls start in one track and end in another. My engineers say that if we build bridges or lay barriers across some of the intersections, we could keep different bulls from ever coming into direct contact with one another. The people would like it and the beastly bull fights would stop."

While Señor Alcatraz was speaking, Evangeline had been busily sketching on a piece of paper.

"Let me make sure I understand," she said, showing us what she had drawn. "Here are the bridges and barriers you allow [see top figure on p. 14]. Here is the original design, where the tracks are labeled *A-A′*, *B-B′*, *C-C′*, and *D-D′* [see bottom figure on p. 14]. Here is your present design with cross-tracks" [see figure on p. 15].

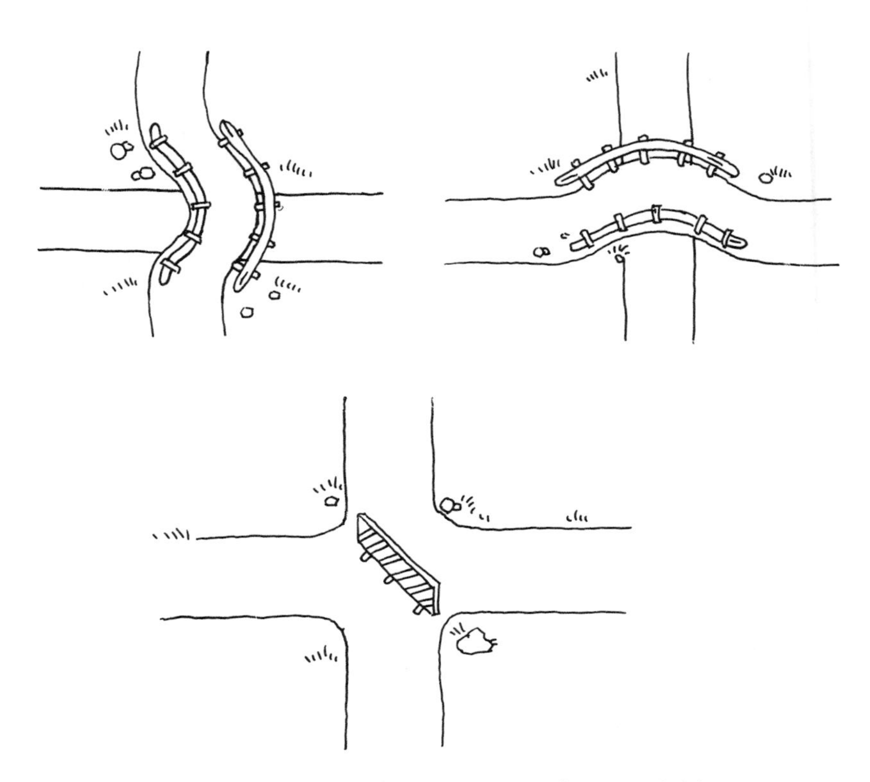

Evangeline's sketches of allowable barriers and bridges.

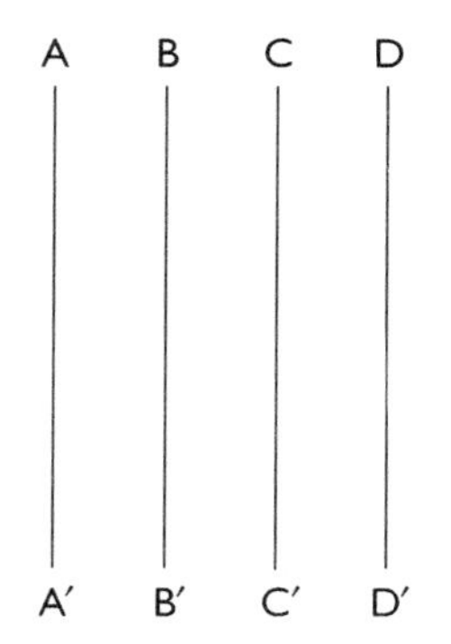

The first design. (Grandstands along the tracks not shown.)

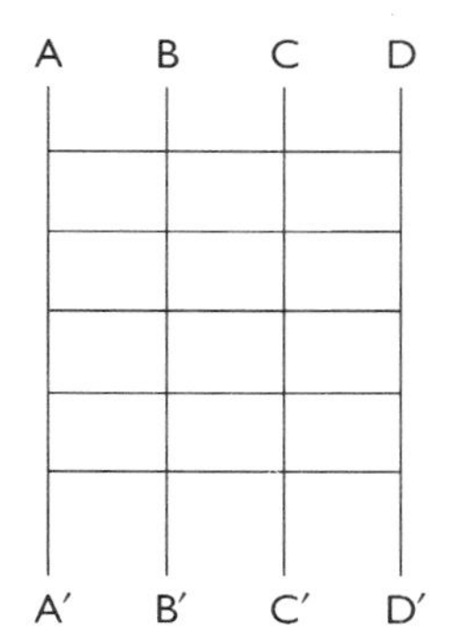

Second design. Five cross-tracks.

Señor Alcatraz nodded.

"You want races in which bulls start at some begin-point X and run to some end-point Y' that is different from X'," Evangeline continued. "Each bull has a fixed route during a race. You never want two bulls to run on the same track or cross-track, because they might fight, but you will allow the routes of two bulls to meet at intersections between tracks and cross-tracks, because there will be either a bridge over the intersection from cross-track to cross-track or from track to track, or a diagonal barrier. Is that right?"

"That's right exactly," said Señor Alcatraz. "Also, the bridges are easily transportable. My men can rearrange all the bridges between races if we so desire. But we haven't yet found a way to design a race with four bulls where crossing tracks is allowed and where routes meet only at intersections. We can do it for two bulls, though. Suppose one starts at A and goes to B′. The other starts at B and goes to A′. Look." Alcatraz quickly drew something on the back of Evangeline's sketch. "This shows how they can run the race without fighting with one another" [see figure on p. 16].

Evangeline nodded. "I understand. Now, you want me to design a four-bull race where crossing tracks is allowed, but only four tracks and five cross-tracks are used?"

"Yes, Señora, that's what I want you to do," Alcatraz answered. "It's even acceptable for some bulls never to use a cross-track, provided at least one bull ends at a different track from the one where it started. Of course, no two bulls should end at the same destination."

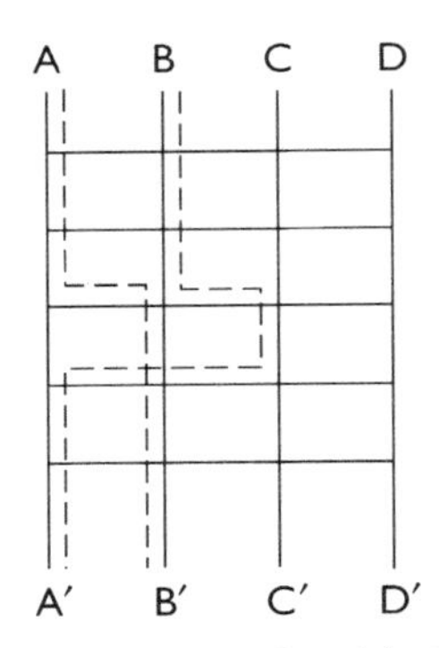

A routing for two bulls. Routes meet only at intersections between tracks and cross-tracks.

? **1.** Is it possible to construct a race with four bulls on four tracks and five cross-tracks, where at least one bull ends on a track other than the one on which it started? What if more cross-tracks are allowed?

"I see," said Alcatraz when he understood the answer. "Suppose I build an extra track from E to E′ and extend the cross-tracks to intersect with E-E′. Both E and E′ are in the shadows, so I do not want a bull to start at E or end at E′; however, I can use the track between them. In fact, could I handle all possible races among four bulls, assuming that I am willing to move the bridges and barriers between races? Do I even need the fifth cross-track?"

? **2.** Is there any pairing of begin-points and end-points (mathematicians call this a permutation) that uses the extra track and requires all five cross-tracks? Or are four cross-tracks enough for every permutation?

"One more question, Dr. Goode: Are three cross-tracks ever enough for four-bull races with an extra track but where each bull ends at a different track from the one where it started?"

? **3.** Is there a permutation where every bull ends at the end-point of a different track than the one from which it started and for which three (or even two) cross-tracks are enough? Remember that no bull should start at E or end at E′.

Señor Alcatraz's eyes lit up as he imagined these races. He mounted his horse, now freshly watered and brushed by the horse-breeder. "Señora, you may have saved many bulls from useless slaughter in the rings. But now let us suppose the people make this a great success and want to see ten bulls racing at once. Assuming I build an eleventh track and am willing to build as many cross-tracks as necessary, will I be able to construct any race I want? What is the minimum number of cross-tracks that I need to construct any race I want? Would I save on cross-tracks if I built more tracks?"

Alcatraz's horse bucked and neighed just as Evangeline answered him, so I never heard the response.

With a bow of his head, Alcatraz thanked us both. "*Hasta luego.* If I can ever be of assistance, please let me know. You have helped me a great deal." With that, he turned his horse away from us, urged it into a slow gallop and was quickly out of sight.

Contest Puzzle 1: What is the minimum number of cross-tracks that Señor Alcatraz needs to be able to organize every possible race consisting of 10 bulls that start at one end of the first 10 tracks and finish at the other end of those 10 tracks? Show that the number of cross-tracks you specify is enough for any such race. Show that there are races for which fewer cross-tracks is too few. Would he be able to use fewer cross-tracks if he had more than 11 tracks?

3. Humpty Dumpty at the Irish Cliffs

We started back to our hotel, tired but eager to decode Ecco's second letter. In town, a newspaper boy thrust the latest edition of *La Prensa* in our faces. Our Spanish was rusty, but we had no trouble translating the headline: "Norteamericanos off Ireland: The Humpty Dumpty Armada?" An accompanying photograph showed three air-craft carriers and scores of supporting ships. The caption read,

"Hundreds of U.S. planes in the Atlantic near the Irish Cliffs of Moher search for the spy ship *Freedom*."

With the help of the newsboy, we translated the article: "The U.S. military apparently discovered three days ago that the U.S.S. *Freedom* was missing but had not announced the disappearance until today, when the presence of a large fleet in the search area made continued secrecy impossible. So far, the U.S. search fleet has been unsuccessful, and there appears to be little chance of recovering the *Freedom* intact. For this reason, the Norteamericano press has come to refer to the fleet as the Humpty Dumpty Armada. . . . "

The newsboy thought we'd be amused at this piece of North American folly, but we were more interested in reading Ecco's second letter. We bought the newspaper anyway and hurried on to the hotel.

I opened the letter and saw immediately that it had been written in a hurry. Ecco's usually sloppy writing was even sloppier than usual. As before, he used a code, and we had to refer to the first letter to see what it might be.

Jkwr K wbj N:

Ox on kwrhf Nwxzrjwf iarbobt wbj O iotmy haoy el xeil yb klipfls ynm jmzzmt zc zon wddm pevro hvoonov. Po avo ef edv paj be Svolafn. Haryovqfego rct ifd tshye hgy uedm. V'p ihe ugyr stn, sdksyf fufg vu sgtu hwzy uvnzyzlm haivc h yxmdj omnqwdzijamn xajy Wnojuwn Emvkmdn, nyoxe nc Xu Pjabon. Dxezkeb az jqpac mbmn vqp mha sqyerj, tpn E zqptn en.

Ca daayd dpdgeseqpdxv cnhhw. "Odc Uzghhf sh Asedk," fe hqzh ltxxivn gji hsxfj hkti nehtkwoh, "uji Afmnnl yn Gykjq. Ezy kngg gntj wlki, Kcca, O wi nzrk."

O wi bax. Dzx pahhal ik xa Orkhwbj. If bkvx hkxxkr lohh dk ob xml Vxjlo jaolaj pc Dbsv. P nygm kylm y ozgra ds hrqrarew ra opofoiov S'm sf bqp bwpb. Kfe qtmm utgo crq ifxcdgx dg jq c tdxchzq jfd tuhiqxq ncpl hcoqp Chhq G'Ighhqnn. Ngicn wqgwnq dsvib uts'v kzegn twfu fqq vub hgre kz Ylmg xhklu mgboyldx yl Yburglt. Njh xwv zy hm lkiyihmuzme evmvhslezyi. Ivss xwd rml idw jdranf jm banv rmld omlzan Aziio M'Omnnwtt. Zyx ybuu gbrx snm kzy poly.

U

"The letter raises more questions than it answers," Evangeline said when we finished deciphering the text. "Did Baskerhound know about the spy ship's disappearance before the fact? Or did the military lie about when it happened? In either case, it seems unlikely that Baskerhound's hurried trip to Ireland is a coincidence. But why would he want a spy ship?"

"More than the letter is puzzling," I said. "Who told Alcatraz who we were? Surely, Ecco wouldn't have."

Evangeline nodded thoughtfully. "I wonder what he's up to," she said under her breath. She was staring at the ceiling. I was about to ask, which "he"—Alcatraz, Baskerhound, Ecco, or somebody else? Before I could speak, Evangeline shook herself out of her reverie and reached for the phone.

"If we're going to find Ecco, we'll have to fly to Ireland," she said as she began to call airlines. She learned that the earliest flight left Montevideo at 2:30 P.M. the next day.

"We have the evening to ourselves, Professor. I propose that we visit El Casino in Punta del Este. Maybe we can learn something by meeting some of Baskerhound's friends."

4. El Casino

We arrived at El Casino dressed in our best clothes, but as soon as we entered we felt as if we were in rags. El Casino was decorated to impress its clients: the chandeliers were Waterford crystal, the carpets dark cognac, the ashtrays silver, and the walls were covered with smoked mirrors from floor to ceiling. The clients were decorated to impress one another: women wore St. Laurent dresses and diamond-studded earrings, men in English suits smoked Dunhill cigarettes through monogrammed filters. The scene was Punta del Este in microcosm.

Punta del Este is geographically part of Uruguay, the southeasternmost point of the country, where the warm Brazilian current mixes with the fresh water of the Rio de la Plata. But nearly everything about this resort city is Argentine: its founders, its visitors, and its allure. Going to "Punta" during the summer season was a sign of success in Argentine society akin to joining an exclusive club in London or being a regular on Rodeo Drive in Los Angeles.

The wealth of Punta seemed immune to the downturns of the Argentine economy. During one economic crisis, the Argentine minister of economics issued an appeal to his wealthy countrymen asking them to take their vacations in Argentine resorts to help save foreign currency. By the time the appeal was published, the minister had gone off on his own holiday—to Punta del Este.

That night El Casino had a new game, and the owner-manager, Donaldo Rumtopo, was demonstrating it to his customers. "Hi-Lo is a game played with two dice, *señores y señoras*," he announced. "A single roller plays against the casino. The roller can bet either 'Hi' corresponding to dice totals of 8 through 12 or 'Lo' corresponding to dice totals of 2 through 6. If the roller rolls a number corresponding to his bet, he wins an amount 25 percent more than he placed."

There was a murmur in the audience. Some were nodding approvingly, others expressing outrage. "You are a thief," said a well-dressed paunchy man surrounded by sinister-looking bodyguards. "I suppose the roller loses on a 7 too. The odds are too poor."

? 1. What would the odds be in that case? Assume fair dice.

"Well, Señor," said Donaldo, looking a little frightened. "We have a variant, if that would please you more. That game is called Winner's Hi-Lo. Every roller starts as a 'slow roller.' As a slow roller, he does indeed lose at 7. However, if he wins, then he becomes a fast roller. If a fast roller rolls a 7, neither he nor the casino loses and he remains a fast roller. However, if a fast roller loses (that is, if he bets 'Hi' and rolls a 'Lo' or bets 'Lo' and rolls a 'Hi'), then he becomes a slow roller again. You can bet as much as you want as a slow roller, but you cannot bet more as a fast roller than the least you ever bet as a slow roller."

There was more murmuring in the crowd. Many people seemed ready to try.

"What are the winnings?" asked the paunchy man.

"A client playing as a slow roller will be paid $1.05 for each dollar placed when he wins. A client playing as a fast roller gets paid $1.25 for each dollar placed when he wins." Rumtopo used dollars instead of pesos for the benefit of the many North American and European gamblers in his clientele.

"Hm, so you might lose less often," I said, "but you get paid less when you win."

The paunchy man turned and walked towards us, his bodyguards following. He bowed politely to Evangeline. Although he had not been introduced, he addressed her by title.

"Well, Señora Doctor, what do you say? Will the casino still have the advantage? Forget about the rolls in which no money changes hands. Think only of the other ones, the 'significant ones.' On the average, how much does the casino win per significant roll?"

? 2. Can you answer the man?

Evangeline showed no surprise at being asked such a question by a perfect stranger. She thought for a minute then gave her answer.

The man nodded and then announced to Donaldo Rumtopo, "I'd be willing to play, if you gave me $1.25 when I win in the slow state and only $1.05 when I win in the fast state." Turning towards Evangeline, he asked, "Am I right, Señora? Will I do better this way?"

"I'm afraid not," Evangeline told him.

? 3. Can you see why she came to this conclusion?

"We will meet again, Dr. Goode," said the man with a polite nod of his head. "I will introduce myself to you then. In the meantime, I wish you and the good professor the best of luck in this house of chance and ostentation."

He walked off, surrounded by his assistants and bodyguards. Evangeline and I stayed behind for some time, but no one else approached us and our attempts at entering into conversations were met with blank stares or polite giggles in response to our incompetent Spanish.

5. Greed and Getaways

We flew to Shannon the next day. The flight was long but bearable, so we rented a car and set off for Moher without stopping to rest. We thought we should investigate the disappearance of the spy ship *Freedom* as quickly as possible.

The many castles and ruins gave the countryside a romantic, medieval character. But I saw only places where a late-twentieth-century villain might hide.

"Professor, something occurred to me when the horse-breeder described Baskerhound's horticultural collections," Evangeline said. "I vaguely remember hearing that at Princeton he had studied chemistry as well as mathematics before going into philosophy."

"It seems possible. Maybe the Director will know," I answered. But Evangeline did not follow-up her remark.

We had reached the coast. A strong westerly breeze had whipped up the ocean, and waves hit the rocks with a fury that sent their foam high in the air. We saw an occasional naval vessel, though we had been told that most of the U.S. fleet had given up the search.

The Cliffs of Moher rise six hundred feet vertically from the Atlantic. The westerly storms, like the one we were now caught in, eroded deep caves into the flagstone. The view from the top of the cliffs was awesome, especially in such weather. There is a strong temptation to climb to the edge of the cliff to get a better view, and each year more than a few tourists die by slipping from the crumbling edge.

Irish soldiers were still searching the cliffs for the missing crewmen, and so most of the paths along the cliffs were off-limits to tourists. What we were permitted to see gave us no clue as to the whereabouts of Ecco or the *Freedom.* Evangeline suggested that we adjourn to a local pub to see what we might learn.

We each ordered a pint of Guinness and found it easy to start a conversation about the recent events. The regulars were only too glad to tell us what they claimed to have seen, especially after we offered to buy a round.

"I saw it with me own eyes," said one. "The *Freedom* went out into the Atlantic."

"No, sunk she did," said another. "Crashed against the cliffs."

Sure such contradictory information would do us no good, I was ready to leave. But Evangeline put her hand lightly on my arm as she glanced at the broad-shouldered man who had just come in.

The man surveyed the bar, and either he noticed our attention or spotted us immediately as outsiders. He walked towards us.

"I am Inspector O'Getman," he said pleasantly. "What might your names be?"

"This is Evangeline Goode and I am Justin Scarlet," I told him.

His eyes lit up. "The philosopher and the mathematician?" he said. "I have just finished studying the Eccoan methods for the last six months. I can use a good dose of them for the case I'm investigating now, I can tell you that . . . but wait, are you here looking for Dr. Ecco himself? Do you think he is mixed up with that crook Baskerhound?" Responding to our surprised looks, he didn't even wait for our answer. "Oh, yes, we know about Baskerhound. How could we not? His fleet of LearJets come regularly to Shannon. He posts his own guards around the planes. At first we thought he was a drug trafficker, but we have found nothing on him, nothing at all. This time, though, we might be lucky. Two of his men were seen near Adare about the time of the bank robbery—what, you haven't heard about it?"

I realized that neither Evangeline nor I had seen a newspaper since leaving Uruguay.

O'Getman went on, "The thieves took gold bars worth nearly $60 million—a good portion of the Republic's reserves. When we heard reports that Baskerhound had been seen here in Moher last Sunday, we decided there might be a connection. But I've found out nothing."

He sat down next to us at the bar and ordered a pint. Evangeline looked at me and shook her head. It was clear she thought we shouldn't tell O'Getman about Ecco's encoded letters. O'Getman downed half of his glass in one go.

"It was a terribly clever robbery," he said after putting down the glass. "But what really upsets me is that the witnesses seem as greedy as the thieves. You see, there were 11 possible escape routes. The robbers had two cars. We know they left by different routes, but we don't know which ones. We have put up a reward for anyone who can name two routes of which the robbers took at least one. Anyone who identifies both routes correctly gets an extra reward. So some of the witnesses who saw one of the cars go down some route just guess the route the other car may have taken. Only a few indicate the one route they are sure about. Is the love of justice completely gone?" O'Getman finished off his beer.

"My question," said the inspector "is this: suppose that everybody is right about at least one of the routes they assert. Then how many different witness reports do we need to figure out the routes the robbers took?"

"You could be asking two different questions," Evangeline replied. "You could mean, What is the maximum number of different answers you might need in order to be sure which routes were taken? Or you could be asking, What is the maximum number of different

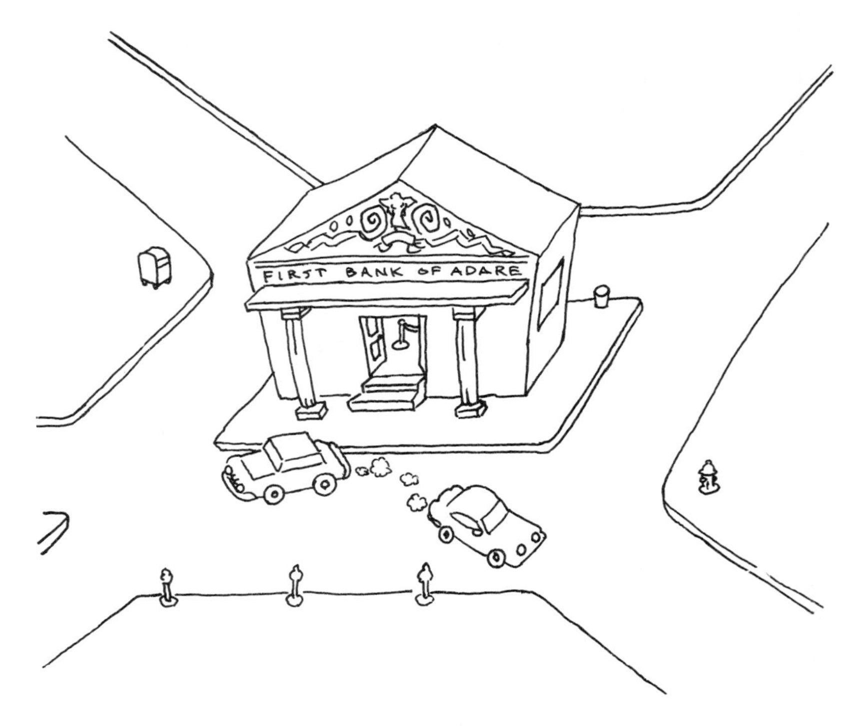

Possible escape routes from the bank in Adare. There were two cars and a lot of greedy witnesses. How many witnesses must O'Getman interview to determine the two escape routes that the thieves took?

reports you would need to present to someone else to demonstrate that your conclusion is correct?"

"Kindly answer both questions," said the inspector.

? Try to answer the two questions. Remember that each witness tells the truth about at least one of the cars.

Evangeline answered his question in only twenty minutes. O'Getman nodded in appreciation. "That's right," he said. "That must be right. Now it is my turn to give you some help, what little I can. A few days before the robbery in Adare, an American took a room in the Adare Hotel—a recently converted mansion with a world-class golf course owned earlier by the Earl of Cumeen. He gave his name as Moley, Kyle Moley.

"Moley aroused no particular suspicions. He was in early middle age, good-looking in a regular sort of way, and he clearly could afford his stay. But he was neither a businessman nor a golfer—unusual in that hotel. And more peculiar, particularly for an American, he liked to take walks. In fact he took walks in Adare each day and late into the evening. Now, you must understand that Adare is a lovely little town, but there is only so much to it. Moley walked through the same streets day after day. He never entered the bank as far as we know and was in a pub when the robbery took place.

"But a few minutes after the thieves made their escape, before even the police were aware that anything was amiss, he put in a long-distance call to an unlisted number near Washington, D.C. One of our telephone operators listened in—they are a nosy lot, I must say."

O'Getman took out his notebook and read from it. "Listen to what he said: 'The hound bit, made off with the brick.' The response came back, 'Good. Return to the room.' That's it. We thought the operator might have misheard, but we had her hearing tested and it's perfect. She also has many American relatives, so her understanding of the American dialect—excuse me, accent—is excellent."

O'Getman's information struck me as more irrelevant than helpful, though the use of the word "hound" may indeed have indicated some connection with Baskerhound. But what good could such a phone call do? Still, Evangeline nodded as if some suspicion has been confirmed.

"Most informative," she said. "Thank you, Inspector O'Getman. And now we'll be pushing on."

O'Getman shook my hand warmly. Then he turned to Evangeline. "Beautiful and brilliant," he said softly as he clasped her hand in both of his. "If you ever need my help when you're in the Republic, please give me a ring."

"Good luck, Inspector," she said with a radiant smile.

"Let's go, Dr. Goode," I said, as I gave her a slight push towards the door.

6. Museum Tour

We drove along the coast south to County Cork. It is no wonder that Ireland is a land of poets. To our left, we saw gentle green hills dotted with farmhouses and mossy castle ruins; to our right, we saw a swirling and furious ocean.

We left the coast to cut across the neck of the Dingle peninsula, then passed through a forested mountain range into County Cork. We arrived in Skibereeen at night. I collapsed into my hotel bed, where I dreamed about a U.S. Navy ship entering a cave and never leaving, perhaps sailing through some secret underground passage . . .

I woke up to a pounding at the door. Evangeline was calling, "Professor Scarlet, let's go. The museum closes at noon today."

The Kames museum was a small house on a hill in farm country, its garden overgrown with bushes and weeds. Anne O'Connell, the curator, clearly paid only the most cursory attention to outward appearances. When we entered the slightly musty building, we were confronted with a huge table, covered mostly with stones.

"These are prehistoric stones," Mrs. O'Connell was telling the museum visitors. That wasn't hard to believe: most stones are prehistoric. She went on to explain the cuts in the stones as evidence of an Inca settlement. Her visitors nodded slowly, unsure whether to agree that this was a major find or to dismiss Mrs. O'Connell as a crackpot.

When the other visitors had left, Evangeline approached the curator and said, "We are looking for our cousin Isaac O'Connell. Could you help us?"

Mrs. O'Connell went to get Ecco's letter, hesitated a moment, and then handed it to Evangeline. Evangeline said, "When did our friend give this to you?"

"Only yesterday morning," said Mrs. O'Connell. "Poor man. He is in bad company; they may be English. My father and uncle both fought the English—back in the 1916 rising. Here is a picture of the independence parade. My father is the man in the second row with his face covered." Mrs. O'Connell paused. She stared intently at Evangeline, who quietly returned her gaze.

"You know," she said. "You look like an O'Connell. We've always looked a little Asian."

Evangeline thanked her politely and we left. We drove back to Skibereen and carefully opened the letter. The task of decryption that we faced looked difficult:

[Jkwr Kew], [Lk wrk h], [kweobt x], [ajwf par], [Wnow, O], [btkhobk], [wbj Ncwr], [hkx, Lm], [wx on Dw], [Iamkr,], [w Z.N. B], [wef nmoq], [jonwqqk], [wrkj. Na], [joj poe], [k ap Dwn], [ukrmazbj], ['n rkxob], [zk -- wh], [h kv-bwe], [wh appoc], [krn. Xmr], [kk iark], [jonwqqkw], [rkj bkwr], [Wjwrk.], [O jab'x], [ubal lm], [wx xmkf], [joj, dzx], [Dwnukrm], [azbj lwn], [ob wb], [zbcmwrwc], [xkronxoc], [whhf taa], [nukrmazb], [j zq xa?], [Xmk bot], [mx lk wr], [roekj ob], [j iaaj x], [j par fa], [zr nkreo], [ckn mwn], [qwnnkj.], [Dzx jab'], [x qwcu f], [azr dwtn], [fkx. Lk], [lohh dk], [ajwf wx], [drkwupwn], [x. "Kcc], [a," mk], [nwoj, "], [xmk kcab], [aioc bkk], [caiqwbo], [abn par], [naik xoi], [kj xa ga], [ob zn."], [O'i wp], [rwoj ap], [mon taaj], [iaajn w], [bj wi la], [rrokj wd], [azx fazr], [nwpkxf,], [if prok], [bjn. Faz], [iwf dk], [dkxxkr a], [k xa cai], [k. Iwfdk], [fazr xl], [a prokbj], [n cwb dk], [qkrnzwj], [pp rkxzr], [bobt mai], [k. Iwfdk], [faz cwb], [cabeobc], [par Wno], [w, O xmo], [bu. Gwc], [ad], [Kewbtkh], [obk wbj], [Ncwrhkx,], [Lmwx o], [n Dwnukr], [mazbj zq], [xa? Xmk], [botmx l], [k wrroek], [j ob Iam], [kr, w Z.], [N. Bwef], [nmoq jon], [wqqkwrkj], [k xmk Jo], [rkcxar x], [a toek f], [az naik], [qraxkcxo], [ab. Lk w], [rk hkweo], [bt xajwf], [. Na joj], [poek ap], [rk. O j], [ab'x uba], [l lmwx x], [mkf joj,], [dzx Dwn], [ukrmazbj], [lwn ob], [wb zbcm], [wrwcxkro], [nxocwhhf], [taaj ia], [aj xajwf], [wx drkw], [upwnx. '], ['Kcca,"], [mk nwoj], [, "xmk], [kcabaioc], [bkkj pa], [r fazr n], [Dwnukrm], [azbj'n r], [kxobzk -], [- whh kv], [-bwewh a], [ppockrn.],

[Xmrkk i], [ark jonw], [qqkwrkj], [bkwr Wjw], [kreockn], [mwn qwnn], [kj. Dzx], [jab'x qw], [cu fazr], [dwtn fkx], [. Lk loh], [wfdk faz], [r xla pr], [okbjn cw], [b dk qkr], [nzwjkj x], [a gaob z], [n." O'], [i wprwoj], [ap mon], [taaj iaa], [jn wbj w], [i larrok], [j wdazx], [fazr nwp], [kxf, if], [prokbjn.], [Faz iwf], [dk dkxx], [kr app r], [kxzrbobt], [h dk cai], [qwboabn], [par naik], [xoik xa], [caik. I], [maik. I], [wfdk faz], [cwb cab], [eobck xm], [k Jorkcx], [ar xa to], [ek faz n], [aik qrax], [kcxoab.], [xmobu.], [Gwcad]

We returned to New York the next day. Evangeline left a message for the Director, who called two days later. She told him that she was convinced that the Adare bank robbery had been committed by Baskerhound and that Baskerhound was heading towards Asia.

"One of our best electronic intelligence spy platforms has disappeared," the Director informed her coldly. "Maybe it's been captured. How can you expect me to be interested in some Irish bank robbery or in some missing Ph.D.? For all I know, he's a monk in some ashram somewhere and these letters are forgeries. It's true that he was useful to us a few times, but he wouldn't be the first egghead who went crazy. Good day, Dr. Goode." He hung up.

Evangeline put down the phone and said with a slight tremble in her voice, "Now I'm very worried about Jacob."

The Genius of Georgetown

7. Musical Messages

The more human the problem,
the more mysterious the mathematics.
From Ecco's notebook, "The Duel"

I invited Evangeline to stay in New York in my guest room. Ecco's letter had filled me with foreboding, and I thought we were better off together than separate. She accepted my offer, but I seldom saw her the next few days. When I did, entering or leaving her room at some strange hour, she didn't say where she had been, where she was going. Some of the messages on our answering machine were in Chinese, but she never discussed them with me and I know no Chinese.

For my part, I spent many hours alone in Greenwich Village coffeehouses brooding over what we had seen and heard. On the fifth day of our return, I was at a table in Café Dante when a man sat down in front of me. He looked familiar, but I couldn't remember how I knew him. He was silent for a moment and I studied him, embarrassed that I couldn't place him.

"You don't recognize me, do you, lovely boy?" he said menacingly in a low Welsh accent. "I am Hanson the treasure-hunter. I am also Halley the regular at Sotheby's. I am also a safecracker and a kidnapper. Does that help?"

It did. "Hanson" had once sought help from Ecco. I alone had met "Halley" after an auction at Sotheby's. I was more than a little alarmed. When a criminal lets you know he is a criminal, he has reason to believe you won't be able to use the information against him.

"My real name is Nigel Williams. I work for Dr. Baskerhound, and we know everything about you and Goode. We should. As Hanson, I put the first bugs in Ecco's apartment. I bet you thought I was pacing around his apartment because I was so upset about the treasure. I was wandering around his apartment placing bugs. And you thought yourself so clever."

"Why are you telling me all this now?" I asked, trying to appear calm.

"I want to bring all you old friends together," he said.

He flicked his wrist and two toughs walked in, pushing Evangeline in front of them. She wasn't hurt and even looked relieved.

"At least we can finally see that Jacob is all right," she whispered as we embraced.

Williams pushed her into a chair and leaned across the table. The goons had taken other seats that enabled them to cut off any escape.

"That's right," said Williams. "Ecco has been solving puzzles for us, but he has somehow passed word on to you. You've tried to follow him and have made yourselves utter nuisances. O'Getman nearly captured us in Ireland and did recover half the gold. Now we're safe, but not rich enough—all because of you. If you had stayed in Ireland another day, we would have nabbed you there. As it happens, your trip will now be a bit longer."

It was quite clear that he wasn't giving us a choice.

Williams stood up. "Now come along, you two," he said. "Dr. Baskerhound wants you both in one piece. He says Ecco won't cooperate unless he knows you're safe, but I wouldn't bloody well mind seeing you off now."

They pushed us out of the restaurant and into a car which drove us to a private airfield where we boarded a private jet. The flight was long and, thanks to the manners of Williams and his pals, unpleasant. He finally deigned to tell us our destination: Bangkok.

When we landed we were again shoved into a car, which drove at a frightening speed to the Gulf of Thailand. A launch carried us to the tiny tourist island of Koh Samui, and a jeep took us from the landing dock to the other end of the island. The jeep sped on the badly paved road and the dust lined my throat with sand.

At last we entered a driveway, passing a sign that said in English and—I assumed—in Thai, "Closed for repairs. Beware of rats and thieves." As we rounded the first bend, we could see a shanty. Scruffy naked kids, some holding knives, ran around the dilapidated building. Around the next bend stood a guard, presumably to prevent anyone bold enough to pass the sign and the kids from going any farther. He waved us on. One more bend brought us to a cul de sac, from which we could see several handsome bungalows and the green, clear Gulf of Thailand.

We were exhausted, sweaty, and thirsty, and amazed to be greeted warmly by a tanned Jacob Ecco. Here he was, at last. After two years of chasing shadows, I had come to think that I would never again see the playful eyes and unruly red hair of my friend. "Welcome to Club Baskerhound," he said as we shook hands. "I am glad to see that you are alive and uninjured in spite of the best intentions of Mr. Williams here." Williams looked on sourly at Evangeline and Ecco as they embraced.

I was relieved to see that Ecco was in good health. Almost too good, I thought. Unlikely as it seemed, I wondered if Ecco was a prisoner or an active participant in "Club Baskerhound."

"Dr. Ecco, please come into the monitor room. Mr. Smartee has called in with a new case," said a young woman with straight auburn hair, stylish eyeglasses, and a serious but friendly expression. She introduced herself to us as Kate Edwards. "Dr. Baskerhound urges your friends to join us," she said.

We went into a dimly lit room equipped with three television monitors and several speakers. Ecco and Kate Edwards took seats at a console; lounging in an armchair was a man I knew must be Baskerhound. Ignoring Evangeline and me, he never took his eyes off Ecco. No one spoke.

The monitors gave different views of Smartee's office in Georgetown, nearly twelve thousand miles away. Smartee sat at a massive oak desk surrounded by acres of marble floor. We could see on the monitors ionic columns in relief along the white walls, rising to a dome forty feet off the ground, and we realized that the "office" was in the ballroom of his mansion. The only pieces of furniture were Smartee's desk, his chair, and three chairs in front of the desk for visitors. We watched as two men were escorted by a butler two-thirds of the way across the ballroom to the chairs facing Smartee.

"Dr. Smartee," said the older man, "call me John Smith. What I am about to tell you is for your ears only. Is that perfectly clear?"

Smartee nodded.

"Good," Smith replied, and stated his case. "Periodically, the Voice of America sends messages to our agents abroad. It is vital that the messages be short so our adversaries can't figure them out.

"Each agent must receive a message consisting of just four bits. A bit is a one or a zero. We encode each bit in a musical phrase. A phrase that descends in pitch will represent a 0 and each phrase that ascends will represent a 1. The problem is that we have nine agents and nine different messages. As I said, too long a song could arouse suspicion. Instead, we will broadcast one song of no more than 14

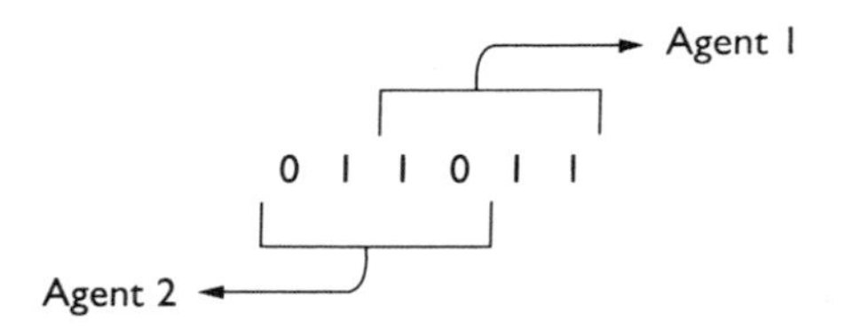

How 011011 would encode the two sequences 1011 and 0110.

phrases such that if each agent starts listening at the right moment, he or she will get the right message. For example, if there were two sequences, 1011 and 0110, then we could encode them into 011011 and tell one spy to start listening at the beginning and the other spy to start listening at position 3. Here are the nine messages: 0000, 0011, 0100, 0110, 0111, 1011, 1100, 1101, and 1110."

I saw Ecco push a button. Smartee shifted in his seat, pushing his finger to his left ear; at the same time he gestured to Smith to pause in his story. "But how will each spy know when to begin listening?" Ecco asked in a whisper.

Ecco turned to me. "The button caused Smartee's seat to vibrate," he said, "alerting him to listen for my question."

On one of the monitors, we saw Smith looking at Smartee with eyebrows raised. Smartee asked Ecco's question.

"We will encode that in the next General Electric advertisement that appears in *Izvestia,*" said Smith.

? 1. Give a sequence of bits of 14 bits or less such that each of the above nine messages is consecutive in that sequence.

Ecco began to sketch out a solution. While he was doing this, Smartee stood up and walked away from his guests through a door into a small room. For the press, Smartee had created the myth that he could arrive at the brilliant solutions that made him famous only in his "Idea Chamber." Our monitors gave us a full view of this fountain of ideas. It was a wood-paneled reading room with a fireplace, a bar, and telematic equipment—fax machine, computer, printer, and a bank of telephones. Smartee went directly to the cognac bottle and poured himself a glass. Then he sat down in a leather reclining chair.

"I hear you've just captured two more brains, Baskerhound," said Smartee aloud. There were microphones in this room, too, I surmised. "With all that gray matter at your disposal, please be quick about it. This so-called Smith is a bore of the first order."

"Here is the solution," said Ecco. "I would be surprised if he didn't have a follow-up question."

Smartee wrote down the solution that Ecco dictated. He preened for a few minutes before the mirror, smoothing his waxed moustache, then returned to the ballroom. He presented the solution to Smith, who smiled.

"You live up to your name, Dr. Smartee," Smith said. "Excuse me, but we knew this answer. The first question was to test you—the Agency is full of skeptics. Now for the real questions. Within twenty-four hours, you must solve three more problems that have duped our mathematicians. First, we may need to send 4-bit messages to five other agents as well. We don't yet know what these messages will be. Can you find a new sequence, 19 bits or shorter in length, into which we can fit any 14 messages?"

Ecco signaled Smartee with the button and whispered, "You mean you can't tell me what the new messages will be, but you want me to give you a sequence that includes every possible set of messages?" Smartee repeated the question to Smith.

"Yes," Smith responded. "I'm afraid so."

? 2. Try to find a 19-bit-long sequence that includes every collection of 14 4-bit messages.

Ecco pressed another button. Immediately, we saw Smartee lean forward. "I understand this problem, Mr. Smith," he said earnestly.

Ecco was already writing down the solution.

"Good," said Smith. "Now, suppose we gave you some particular set of 14 four-bit messages. Can you encode them into 18 bits?"

? 3. Notice the difference between this problem and the previous one. In the previous one, you were asked to find a sequence of 19 bits into which any 14 4-bit messages could be encoded. In this one, you will be given some particular set of 14 4-bit messages and you are to find an 18-bit sequence that encodes them. You must show that you can do this no matter which set of 14 4-bit messages you are given.

"Can you do better than 18 bits for every given set of 14 4-bit messages?"

? 4. That is, can you prove, for some set of 14 4-bit messages, that no sequence of 17 bits or *fewer* can encode them all? Alternatively, can you show that a length-17-bit sequence is enough for any possible set of 14 4-bit messages?

In spite of herself, Evangeline became engaged in the puzzle and solved the fourth problem at about the same time that Ecco answered the third. Kate Edwards gave these solutions to Smartee, who wrote them down—between sips of cognac. It occurred to me that it was early morning in Washington.

8. Elves Flip

A gong sounded. Baskerhound stood up and motioned for Ecco, Evangeline, and me to leave the monitor room with him. Kate Edwards stayed behind to watch Smartee deliver the solutions.

"Welcome, Professor Scarlet and Dr. Goode," said Baskerhound, jovial now. He was a tall, somewhat heavy man with brushed-back, thick, wavy brown hair. His large blue eyes had an irrepressible sparkle. "I'm delighted to have you as my guests—I shall be pleased to introduce you to our island's finest restaurant."

He shook my hand vigorously and tried to kiss Evangeline's, but she drew it back. He smiled slightly and escorted us to golf carts that took us across the grounds to what appeared to be a charming country restaurant overlooking the emerald gulf.

"This table is reserved for us," he said, indicating a table for four in the corner of the dining room nearest the sea. "All the people you see here are my staff." A gentle breeze brushed the scent of spices and the smell of steaming rice in my direction, and I realized how hungry I was.

"You have joined an odyssey for freedom," Baskerhound went on, as we helped ourselves to the food that had been placed in the center of the table. "We seek freedom from state, religious, and corporate control."

"Holding us prisoner is evidence of your dedication to liberty, no doubt," Evangeline said with ill-disguised contempt.

"My dear Dr. Goode," said Baskerhound, "just as an omelet requires broken eggs, so does my effort require me to compel the services of Dr. Ecco. He, in turn, needed to know that you were safe. So, you both are here. It is a sequence of simple syllogisms."

Evangeline said nothing.

Baskerhound stared at her in evident admiration, then turned to Ecco. "Donaldo Rumtopo sent a letter that arrived today. He wants me to invent a game for El Casino that uses a new coin-flipping machine recently introduced by Bally in Atlantic City. It flips the coin — Bally recommends a one-ounce gold American eagle — by a jet of air whose force the player can partially control, collects the coin after it lands, and funnels it back to the air jet. It counts the number of heads and number of tails since the last time the machine was reset."

"What kind of game do they want?" asked Ecco.

"One in which a player gambles against the casino while spectators may bet with the player," Baskerhound answered. "Since the casino prefers that patrons have frequent occasions to wager, a single game should last no more than 10 to 15 flips. The game I devised is called 'Elves Flip.' A player can choose either heads or tails and flips a coin 11 times. For the sake of argument, assume the player chooses heads. The player wins if heads wins throughout the game. That is, the casino pays the player $7.00 for every dollar wagered, if, after each toss, heads appear more than tails. Otherwise the player loses the bet. Ecco, seven sounds like a good number for a game of chance, but will the casino win on the average?"

"Well," said Ecco, "to be ahead all the time, the player must win the first and second toss. But that doesn't get us very far."

"I have an idea," I said. Like Evangeline before, I couldn't resist the challenge. "Suppose we consider each possible final outcome one at a time. That is, we will consider first that the player gets six heads and five tails, then the outcome of seven heads and four tails, and so on. For each such outcome, we compute the likelihood that there will be more heads than tails from the first toss to the last. Since at most one such outcome is possible, we can just add up those likelihoods."

"Nice idea," said Ecco—high praise from him. "Well, how about it, Baskerhound? We will start by computing the likelihood that the number of heads is greater than the number of tails from the first flip to the last for some particular final outcome, say seven heads and four tails. Then we'll see whether we can generalize the results."

"Not a bad approach, not at all bad," Baskerhound said. "But even that isn't so simple."

"Now, Professor, perhaps you could provide me with certain bits of information," Ecco said.

"I will try," I answered.

"Here are my questions," he said. "You may wonder at their relevance, but if you think about it, you will see.

1. What is the likelihood that heads will win with a final outcome of seven heads (whether or not tails is ahead sometime during the game)?
2. What is the likelihood that heads will win with a final outcome of seven heads and that the first flip shows tails?

"You should have no trouble," Ecco assured me.

? Note to readers who have not studied probability: Professor Scarlet takes care of all the calculations in this puzzle. The flash of brilliant insight is left to you. Readers who do know probability are asked to help Professor Scarlet out.

"Well, let's see," I said. "There are $2^{11} = 2048$ sequences of flips. All of them are equally likely. The number of ways seven heads can be distributed among eleven flips is $\binom{11}{7}$ or 330. So, the likelihood of that outcome is 330/2048. Similarly, the likelihood that there will be seven heads among the last ten tosses is $\binom{10}{7} \times 1/(2^{10})$. However, even in that event, the first toss may be heads or tails. So the answer to the second question is 60/1024. "Not that I see how that helps, Ecco. The first answer is the probability that heads will win by three and the second is the probability that heads will win by three after losing the first flip. So what?"

"Quite right," said Ecco, "but you must think about reflections. You have given me all the information, I need."

? 1. What is the likelihood that the outcome is seven heads to four tails and that heads will lead from the first flip on? Ecco's comment is a big hint.

"Ecco, that is smart," said Baskerhound after hearing Ecco's solution to that special case. "Determining the odds for the game as a whole is now just a matter of calculation."

Ecco looked at me. I already had the appropriate information ready for the other possible situations in which heads wins.

1. The likelihood that heads wins 6 to 5 is 462/2048. The likelihood that heads wins 6 to 5 and that the first flip shows tails is 210/2048.
2. The likelihood that heads wins 8 to 3 is 165/2048. The likelihood that heads wins 8 to 3 and that tails is the first flipped is 45/2048.
3. A 9-to-2 win for heads occurs with probability 55/2048. A 9-to-2 win and a loss on the first flip occurs with probability 10/2048.
4. A 10-to-1 win for heads occurs with probability 11/2048. A 10-to-1 win and a loss on the first toss occurs with probability 1/2048.
5. An 11-to-0 win occurs with probability 1/2048. There can be no 11-to-0 win with a loss on the first flip.

? 2. Will the casino like the odds for Baskerhound's game?

After a quick calculation, Ecco presented the solution to Baskerhound. Baskerhound asked a few questions, then smiled with great satisfaction. He snapped his fingers and a bottle of champagne quickly appeared. When all our glasses were filled, Baskerhound raised his glass and said, "A toast to you, Dr. Ecco, and your two able friends. We are on the same side. You'll see."

Ecco and I lifted our glasses in uncertain response, but Evangeline stared sullenly at the table.

Who has better odds, the player or the casino?

9. A Problem of Protocol

The next few days were quiet. Evangeline and Ecco took daily rides on the windsurfers that Baskerhound had left at their disposal, high-performance Mistrals no less. I watched them with binoculars and saw that their long routes were constrained by a network of steel mesh fences that rose above the sea level.

I asked Kate Edwards about them. "Oh, the fences mark the beginning of the labyrinth of sharks," she said matter-of-factly. "Your friends are completely safe on this side of the fence. Beyond it, they are likely to be attacked by the great whites that Dr. Baskerhound keeps as pets."

Ms. Edwards was a delightful companion. Although I recognized that anything I said might be reported to Baskerhound, I felt some reason to trust her. She spoke little about herself, saying only that she was from Surrey and had joined Baskerhound two years earlier to maintain the electronics needed to communicate with Smartee. She considered herself a technician and seemed to know little about Baskerhound's goals and principles, although she knew quite a bit about his often dubious methods.

"Take this place, for example," she said. "A few years ago it was a thriving inn here on Koh Samui. After traveling around the world for months, sometimes years, young Europeans and Australians would come to this island and decide that it was paradise. It is paradise, if you like to live on $5.00 a day, sit on the beach, and occasionally order a special dish of sautéed hallucinogenic mushrooms. Mind you," she went on, "I have nothing against these Euro-waifs. They do no harm to anybody, mostly because they do nothing at all. But Dr. Baskerhound needed privacy. So he bought this inn and the land around it for an exorbitant fee. Knowing he would arouse suspicion and curiosity if he simply booted out all the guests, he mixed traces of arsenic into the food and served only ordinary mushrooms. A few of the waifs became ill and the rest left without further ado. To discourage other visitors, Dr. Baskerhound imported a family of thugs from the underworld of Bangkok and posted a guard around the bend. I understand the word among travelers is that this inn is secretly a toxic waste dump. We have not been bothered since."

Kate—as I now thought of her—was very informative about the island's fauna and flora, showing me the plants that grew along the shore. She pointed out the plants that Baskerhound had transported from the pirate's arboretum in Punta Ballena. Many of them, she told me, were poisonous to eat.

A piercing whistle cut off our conversation. "That will be Mr. Smartee requesting our help," she said. We hurried back to the bungalows and saw that Ecco and Evangeline were already there.

Baskerhound came out of the monitor room shaking his head. "It's the United Nations again. Ever since we settled the war in Indonesia, they come crying to us whenever there is a crisis. Now they are trying to settle the Toocie War between the Rukatis and the Taraks. If you remember, the war started because each country accused the other of being the source of the dreaded toocie virus, a virus that, once contracted, changes one's skin color to a bright magenta after only a few days. These people hate one another so much, they refuse to shake hands. Not that I blame them — the virus spreads from even the most casual contact."

From the monitor room we watched Smartee sitting behind his desk as he faced a tall African dressed in white robes. "Dr. Smartee," said the visitor, "my name is Uwabi, protocol officer at the United Nations. The Secretary General is convinced he can bring an end to this pointless war, if he can just solve what amounts to a problem of protocol. You see, each nation has sent two representatives to a special meeting. There is also a fifth person, an interpreter. The five visitors must shake hands with the Secretary General. In addition, the chief representative of the Rukatis must shake hands with the chief representative of the Taraks. Their assistants must also shake hands with one another, but protocol forbids an assistant to shake hands with a chief representative — these are both very hierarchical cultures.

"Everyone, including the Secretary General, is worried about catching 'toocies.' I mean everybody. The Secretary General is worried. Each Rukati is worried about catching toocies from his fellow Rukati. The same applied to the Taraks. Since touching one bare hand to another bare hand would transmit toocies back and forth, they want to use special gloves during the meeting ritual. Unfortunately, only three such gloves are available. Their properties are:

1. The inside and outside of a glove can both carry toocies. However, the glove is impermeable to toocies, so toocies won't travel from one surface to the other.
2. If any hand or glove surface touches a hand or glove surface, then each touched surface acquires all the toocies of both surfaces. That is, toocies are exchanged and shared.
3. A glove can be turned inside out — using tweezers — without spreading toocies. Tweezers can also be used to pull one glove over another or to pull one glove off another. You can assume that the tweezers will be washed with chlorine after each use. That treatment will kill the toocie virus. Unfortunately, the gloves cannot be cleaned.
4. The gloves are thin and tight-fitting so a person can put on three gloves at once and shake hands properly.

"The question is: how can the hand-shaking goal be achieved without having anyone touch any toocies but his own? For example, suppose there were three visitors who had to shake hands with the Secretary General (who must remain bare-handed), but only two available gloves. How would you have them do it?"

? Give this easier example a try.

Smartee raised his finger, Uwabi nodded. Smartee stood and walked to the Idea Chamber.

"This whole affair is frightfully distasteful, Baskerhound," he said. "Please hurry and find a solution. I wonder if cognac kills toocies." With that, he opened the bottle and refilled his glass.

Ecco worked on the problem for quite some time before arriving at an answer. Kate Edwards dictated the solution to Smartee via microphone. She told him, "The first visitor puts both gloves (using

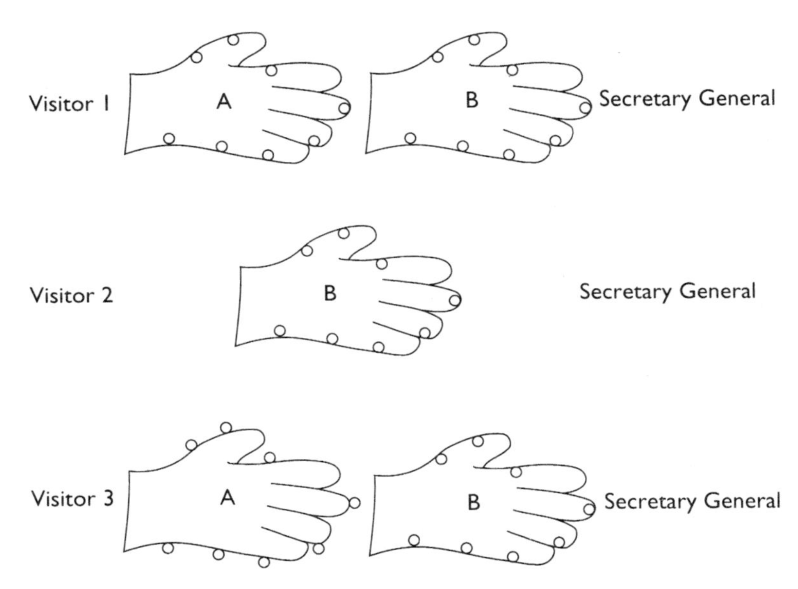

A highly schematized illustration of Ecco's solution to the practice problem. Each glove has two sides: one is represented as having circles and the other is represented as having none. Toocies cannot penetrate from one side to the other.

tweezers), one over the other, on this right hand. He then shakes hands with the Secretary General. Call the inner glove A and the outer glove B. The second visitor uses the outer glove B to shake hands with the Secretary General. (The second visitor gets no toocies, since the inner surface of B has not been contaminated.) The third visitor reverses glove A and puts it inside glove B to shake hands with the Secretary General. Thus, the Secretary General shakes hands with only one surface of B. The other surface of B is used by the second visitor. The first and third visitors use the two surfaces of A."

When Smartee presented this solution to Uwabi, the man made a deep bow. "Well done. The best mathematicians at the United Nations took a week to solve that example. They have given up completely on the problem involving five visitors and three gloves. But we must solve that problem to bring peace to these warring peoples. Can you help?"

Again, Smartee raised his finger, and again, Uwabi nodded with respect. Smartee walked back to the Idea Chamber, closed the door behind him, and sat back in his reclining chair, the cognac and pipe near at hand.

Evangeline and Ecco began an excited conversation. After a few minutes, they had a solution. Since it was a long one, Kate Edwards sent it by electronic mail to the computer in the Idea Chamber. Then she spoke into the microphone: "Print it please, Mr. Smartee."

Smartee touched a button and the laser printer disgorged the electronic mail. After reading the results, he walked with dignity back into the office, carrying his pipe but leaving the cognac behind. "Mr. Uwabi," he said, "here is one solution. I wish you the best of luck in making peace."

Uwabi bowed and left.

10. Tropical Antarctica

By our ninth day on the island, I fell into a comfortable routine of swimming, walking on the beach alone or with Kate Edwards, reading in Baskerhound's excellent personal library, and dining on sumptuous meals. It was clear that Baskerhound's only demands were to solve puzzles and to forget about escaping. Escape was out of the question in any case. Armed guards accompanied by nasty dogs met me whenever I walked more than a few hundred meters in any direction. The sharks precluded any escape by sea.

So we were prisoners, but comfortable prisoners. It still troubled me that Ecco was willing to help Baskerhound. Finally, I asked him for an explanation.

"There are four true reasons, and one of them is the real one," Ecco replied. "First, I'm a puzzle junkie. Smartee and Baskerhound find good challenging puzzles and I feel compelled to attempt them. Second, I have nothing against Smartee's clients. They put my solutions to good purposes. Third, I have no choice. You've met Williams. He would be only too pleased to cut our throats. And fourth, Baskerhound is up to something. I want to find out what it is."

I went away disappointed. I had expected more rebelliousness from my friend, but I agreed that his dying a hero's death would accomplish nothing. And he might have just helped settle a war.

By this time, my walks with Kate Edwards were a daily occurrence. "Baskerhound is quite smitten with Dr. Goode," she said during one such ramble.

"Every man who admires intelligence in a beautiful woman feels the same way, Kate," I replied. "Including me. But she loves only Ecco. And he loves her, in his own fashion."

"I know what you mean," she said. "When Ecco wrestles with a problem, he is consumed by it until it is solved. It's as if his intelligence and insight come from another world at those moments. This world and the people in it matter only for their contribution to a solution."

"Or for their ability to ply him with cookies," I said. We both laughed.

On the evening of the ninth day, the whistle summoned us to the monitor room. The screens showed Smartee with an English military man. "My name is General Jeffrey Falcon Scott, great-grandson of Robert Falcon Scott, the explorer," he said as they shook hands. "Fellow member of Pop, I understand. By the looks of it, you were still in the pram when I was secretary. Ah, Eton . . . a fine group of lads we were back then. Now, so many of us are bankers or bureaucrats . . . "

Smartee smiled but did not answer. It was obvious he enjoyed meeting this client. Scott's mustache was trimmed in a bushy Victorian style, and his entire bearing was that of a soldier of the Empire.

Scott settled comfortably into the chair facing Smartee's desk. "I need your help with a small architectural project," he said. "Recently I read of Dr. Ecco's work on the design of an Antarctic research station. I have a similar problem and similar constraints, but I need more rooms. But, then, I hear you're better than Ecco ever was."

"I haven't had such bad luck up to now," Smartee replied.

I glanced at Ecco. He smiled at me and whispered, "Let him enjoy his fifteen minutes of fame."

Scott continued. "Our rooms are square and all the same size. Their precise dimensions are a British government secret, I regret to tell you; I can say they're really quite large, especially in relation to the width of a standard doorframe. Each room may have up to four doorways, which may be cut anywhere along the walls except at the corners. The final structure should consist of 41 of these rooms in a single-story structure—no hallways or other structural features. It must be possible to walk from any room to any other one through six or fewer doorways. To put it another way, it should be possible to move from any room A to any room B by walking through five or fewer intermediate rooms."

? Try to produce such a design.

"Return around this hour tomorrow, won't you?" said Smartee.

"Right," said Scott. He rose, shook hands, turned, and left.

Ecco provided Smartee with a solution in a few hours. Scott returned the next day in cricket flannels. Smartee handed him the solution.

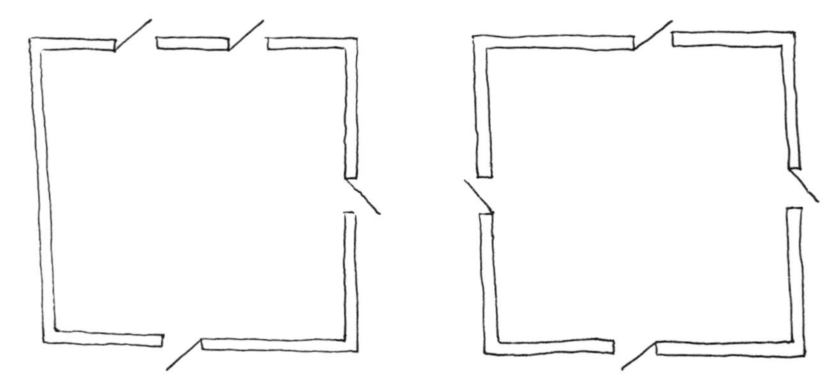

Square rooms with up to four (negligibly narrow) doorways. The doors can be arranged in different ways in different rooms and can be placed anywhere, except at corners.

"Jolly good," said Scott after he had studied the diagram. "Come and join us for some cricket."

"I'll be at the pavilion within the hour," Smartee answered.

Contest Puzzle 2: Suppose that, instead of square, you could make the rooms rectangular with any length-to-width ratio you like, but all the rectangles must have the same dimensions. Otherwise the rules are the same: four doors per room anywhere but in the corners, and it should be possible to travel from any room to any other going through at most six doorways. The rectangles can be in any orientation.

Can you construct such a single-story structure with more than 50 rooms? If so, construct one. Can you construct a structure with more than 60 rooms? If so, construct one. If the answer to either question is no, just say so (no proof is necessary).

11. 100-Day Rockets

Whether to convince us of the virtues of his ultimate goal or to woo Evangeline to his side, Baskerhound spoke with a zeal that would be judged evangelical: "In the past, religion would assert control at all crucial moments in our lives: birth, coming of age, marriage, and death. It gave ritualized comfort for those transitions. In the modern era, the state asserts its control many times a day—every phone call is taxed, street corners are watched. It gives some material comfort to those in need, but at what price? The church created inquisitions and the state constructs weapons for mass destruction. I would have a country with neither church nor state."

And so on. Even admitting his argument, I could not see how it could have more than theoretical interest. After all, states had partitioned the habitable earth, and I did not see them voluntarily withering away.

When I raised my objection, Baskerhound said, "You dismiss me as a theorist, Professor, as if theorists are no better than dreamers. But everyone is a theorist. Everybody lives his or her life based on some principle and some perception of facts. So-called pragmatists follow the principles of the current power elite and consider the status quo to be the only significant fact. I follow the principle of an environmentalist anarchism and view the death of liberty and the destruction of the environment as the significant facts. I hope to change the facts to conform to my principles."

Brave words, I thought. I was starting to ask him to be more specific when the whistle shrieked.

We convened in the monitor room. Smartee's visitor was dressed in a business suit, but he had a U.S. Air Force medal pinned above his left breast. He got right down to the matter at hand.

"My name is Robert Strode," he announced. "Ever since the tragic space shuttle accident of the mid-1980s, the U.S. government has allowed private companies to manufacture and launch rockets. My company, Compact Rockets, has found a wonderfully simple design for small but powerful rockets. The design consists of 100 independent components that can be built in different factories. We like the number 100, so we have decided to build 100 rockets to start with.

What the trucks should accomplish. The arrow indicates the desired redistribution of parts by 100 trucks.

Currently, each of our 100 factories builds 100 copies of a single component. We want to redistribute the copies so that each factory will have one copy of every component. After that, we will assemble the components into a rocket. Unfortunately, building the components is taking longer than we originally expected. If we spend too much time redistributing all the component copies, our 100 rockets won't be ready by our 100th day of business. Both our investors and sponsors might lose confidence in us. We have hired 100 trucks. Each truck can carry 100 component copies in any combination (all the components are small and roughly the same size). The trucks must travel slowly because the components are fragile. We have calculated that each trip between factories will take 5 hours, no matter how many component copies are loaded and/or unloaded. It also takes 5 hours for the trucks to arrive at any of the factories from the truck depot."

Well then, it is best if each truck loads up as much as possible whenever it has the opportunity, I thought.

Strode seemed to read my mind. "At first, we thought it would be best to load as much into each truck as possible. But if a truck takes, say, all 100 copies of component A from one factory, it must make 99 trips to other factories in order to deposit those copies. That would take 500 hours all told. So, we must use some other method. The question is: Is there a method to do the entire job within 100 hours, assuming the trucks never stop? By the way, two trucks can be loading or unloading at each factory at a time."

? 1. Can you do it with 100 trucks in 100 hours?

Ecco solved the problem in a few hours and sent the solution to Smartee. He attached a variant of his solution with this note: "It can be done in 95 hours with this many trucks."

"Mr. Strode, the task is indeed possible," said Smartee. "I can even do it in 95 hours."

? 2. Can you see how to accomplish what Ecco claims?

Strode was, of course, interested. Smartee handed him the solution and told him the fee. Like most of Smartee's clients, Strode did not protest. He immediately wrote out a check for $75,000.00.

"You have saved us much more than that," he said gratefully.

Ecco turned to me and asked, "How many trucks would he need to finish the job in 40 hours? Assume he has 20 loading docks per factory."

? 3. What do you think?

12. Three-Finger Shooting

"The Taraks and the Rukatis have made peace, although intelligence reports indicate that generals on both sides resisted the armistice," Kate Edwards reported after listening to the BBC World News Service one morning. "It seems that both general staffs were convinced that they were on the verge of victory."

"The only certainty," said Evangeline, shaking her head, "is that more people would have died."

"Chess is the wrong metaphor even for two-person zero-sum games," Baskerhound said. Rules of conversation suggested that his remark was connected with Evangeline's assertion, though I didn't see how. "It suggests that both sides have full information and that deliberation is better than unpredictability. But unpredictability is more important in most things. By the way, have you seen my new game?"

With that, he stood up, turned away, and started muttering. "One, two, three, shoot," he said as he extended the index finger of his right hand. "One, two, three, shoot," he repeated, extending his middle three fingers.

"Is that legal?" I asked.

"A new twist on an old game," said Baskerhound with a smile. "I am examining its possibilities for Rumtopo. It seems that the ladies want a low technology game that allows them to show off their jewelry at El Casino. Here's how it goes: in unison, two people — one from the casino and one a player — say 'One, two, three, shoot!' and then each extends either one, two, or three fingers. If the total number of fingers is even, then the casino wins. Otherwise, the player wins. Kids play the game with just one or two fingers, but using three adds a whole new dimension."

"Oh, yes," I said. "If both players choose each number of fingers with equal likelihood and independently, then the casino wins five times out of nine."

? 1. Confirm this.

Each player can put out one finger, two fingers, or three fingers.

"Right," said Baskerhound. "However, if the casino announces that as its strategy, the player can win two thirds of the time."

? 2. How would that happen?

"What exactly do you mean by a strategy?" asked Ecco.

"A strategy assigns a likelihood or probability to the choice of each number of fingers," Baskerhound responded. "For example, the equi-likelihood strategy assigns probability 1/3 to each number of fingers. Are you ready for my questions?"

Ecco nodded.

"Here they are," said Baskerhound.

? 3. Is there any strategy that the casino can announce and still win on the average? If so, what is that strategy?

? 4. If neither side announces a strategy, but both play as well as they can, who should win? What should the casino do and what should the player do? Is it ever good to change strategies?

? 5. Consider a game in which the player wins whenever the total is 4 or 5, but the casino wins when the total is 2, 3, or 6. Which side should win then, assuming they both play rationally?

Ecco and Evangeline teamed up and answered these questions before we finished breakfast. Baskerhound was astonished and not particularly pleased with their results.

"The casino will never like this," he said, frowning.

"Maybe things aren't so bad," I suggested. "Perhaps players will be willing to pay a slight commission if you let them win on evens and the casino on odds."

Baskerhound beamed. "Not at all bad, Professor. Not at all bad."

He leaned back in his chair and looked up at the ceiling. A gecko snatched a large insect from the rafters. This seemed to upset Baskerhound. "Pack your bags," he said abruptly. "We're leaving in two hours."

Two hours later, exactly on schedule, Baskerhound's entire staff, except the cooks, took off in a fleet of jeeps. Then to a launch, then to a car, then to the airport. Within four hours we were in the air, heading southeast. The frown did not leave Baskerhound's brow until the plane left the ground.

Justifying the Means

13. Sand Magic

We in the missile systems division
create aircraft that make one-way flights.
Technical manager from Martin Marietta
to a group of mathematicians

The long plane ride took us across the Pacific, over the Andes, and into Punta Ballena. We left the cramped plane and stretched our limbs, happy to be in the warm, dry noonday Uruguayan sun once again.

Another fleet of jeeps took us to Baskerhound's mansion, concealed in some eighty mostly forested acres surrounded by hedges. Baskerhound showed us to our rooms and said, "Take a three-hour siesta. You'll need it. We will undoubtedly have a late night."

We needed no other encouragement. Some hours later I awakened to a steady drumbeat accompanying a Portuguese song. I dressed quickly and followed the sound of the music to the pool area. The Brazilian band, consisting of steel drums, guitars, singers, and dancers, was regaling Baskerhound, who smiled and clapped his hands. Two dressed lambs turned slowly on a spit over an open fire.

"Welcome, Professor," said Baskerhound gesturing for me to join the party. "We have been working too hard," he said. "The people who know best how to celebrate are Brazilians. That's why you hear Portuguese now instead of Spanish."

I was light-headed by the time dinner was served, but completely relaxed. We all ate the lamb with great enjoyment. Even Evangeline for once relaxed her serious demeanor.

After dinner, Baskerhound ushered us into his private theater, where we were greeted by a troupe of magicians. An assistant brought out an empty dark wooden bucket. He filled it with water while the magician, a short fat man in an elaborately tailored crimsom silk outfit, supervised.

The magician then requested that some member of the audience hand him a coin. I took one out of my pocket, marked it, and gave it to him. His eyes widened. "Why, this is a jumping coin, sir," he exclaimed.

"What is a jumping coin?" I asked.

"Allow me to demonstrate." He carefully placed the coin on the bottom of the bucket, wetting his silken sleeve in the process. Then he signaled the band to resume playing. During a drum crescendo, the coin jumped out of the water. I checked and saw that it was the coin I had marked.

We all applauded. I looked at Ecco and saw that he was deep in thought. A few seconds later, his face lit up in a smile.

"What is it, Ecco?" asked Baskerhound. "Surely you don't doubt the coin's magical properties."

"Magic there is," Ecco answered with a chuckle. "But it is all in the bucket."

"Prove it," said Baskerhound.

Ecco walked onto the stage and poured the water from the bucket.

? Before you read on, what do you suppose fell out of the bucket besides water?

"Ecco picked up a piece of metal in the form of a spiral. "Just as I expected: a spring," he said. "When he placed the professor's coin in the bucket, the magician coiled and lodged this spring with a piece of salt. He put the coin on top of the spring. When the salt dissolved, the coin popped out."

"Bravo, Ecco," said Baskerhound. "Now, please prepare for your real test." He pointed towards the stage and announced, "Ladies and gentlemen: the Amazing Sand Counter!"

A man dressed like a Cossack with high boots and a gilded hat shaped like a mushroom stepped forward. He carried a second bucket. He bowed with a flourish.

"I am the Amazing Sand Counter," he said. "If you put sand into this bucket, I know at a glance how many grains there are," he said. "But I won't tell you."

"Why should we believe you, then?" I asked.

Baskerhound stood up and turned to us. "That is precisely the question. Can you devise a test by which the Sand Counter can prove his skill without telling you anything that you don't already know?"

The Amazing Sand Counter claims to know how many grains of sand there are in a bucket just by looking at it. How can you put him to the test?

"Wait—I don't understand," I said. "You want an experiment that follows this form: I put a large amount of sand into the bucket, more than I could possibly count. So I don't know how many grains there are and neither does the Sand Counter—unless he really has this amazing power. Then I conduct an experiment, which may entail asking him to leave the stage or turn away. At the end, I am convinced he knows how many grains of sand are in the bucket, even though he hasn't told me. Also, whatever he tells me, I have learned during the course of the experiment."

"That's right," said Baskerhound, "Any reasonable person should be able to do this test and be convinced."

? 1. Try to design such an experiment.

Ecco succeeded after about fifteen minutes. When I heard the design, I marveled at its simplicity. But Baskerhound wanted more.

"Ecco, the story does not end there," he said. "Suppose I conduct your experiment and determine that there is at least one Amazing Sand Counter in the world. Suppose someone else, say you, Professor Scarlet, claims to have this remarkable skill as well. Now, I will try to test you using Ecco's experimental design, except that I will allow you the privilege of removing the sand bucket from the experimental room for inspection whenever you want. Could Scarlet pass the test, even without the Sand Counter's talent?"

? 2. Now suppose that a single Amazing Sand Counter did exist. Is there something Professor Scarlet could do to impersonate his abilities? (Hint: The professor may test the Sand Counter.)

"Yes, I think so," said Ecco. "Here is how . . . "

"Baskerhound listened intently to the explanation and then exclaimed, "Of course! That's it! Ecco, you don't know it, but you have changed the course of history." I had no idea what he meant.

14. Signals and Echoes

From what we could read in the U.S. newspapers that came in daily by air, Dr. Smartee's fame had grown steadily. In Washington, power drives the social scene. If you have power, you are invited everywhere. If you don't, you stay home and wait for the phone to ring. The society pages of the Washington tabloids often showed Smartee accompanied by beautiful actresses, heads of corporations, and foreign dignitaries.

"He has class, he has power, and he has brains," wrote syndicated society columnist Susanne Post. "And the ladies adore him."

Serious commentators also recognized his prominence as a puzzle solver who influenced world events. "The lionization of Dr. Smartee is without precedent in the nation's history. Never before has raw intelligence led so directly to power," intoned Ted Bickley, the famous conservative television commentator and personality.

High-ranking military officers began to approach with their most secret problems. Most prominent was Admiral Trober, the former football star now in charge of all U.S. nuclear submarines. We watched him present a new problem in Smartee's ballroom.

"Dr. Smartee," the admiral said, "you may know that our submarines can travel in any ocean on earth and still launch missiles that will hit within a few meters of their targets. Many years ago, we figured out a method for sending messages to these submarines. The method involves transmitting extremely low frequency signals from huge antennae in Wisconsin. The earth and water are our transmission media. A few years ago, our scientists invented a new method that gives us five possible 'tones' as we like to call them: A, B, C, D, and E. Don't ask me for more details—I couldn't tell them to you, even if I understood them. In principle, these tones work better than Morse code, because the earth's echoes can cause dots to reverberate like dashes, causing much confusion."

The admiral paused and took a breath. "However, last year something extremely frightening happened," he said. "The message BCCAD, meaning 'all is well,' was transmitted and one of our submarines interpreted it as CCDBE. According to the code used at the time, that was one symbol off from the combination of tones that means 'prepare for war.' We have done experiments since then and have found that transmitting A may result in the receipt of A or B; transmitting B may result in B or C; C may result in C or D; D may result in D or E; however, E will always be received as E. We represent these possible misinterpretations by a graph. Now, it happens that the tone we tend to send the most by far is B. The other tones occur with about equal frequency, except that E is used very seldom. Our communication specialists suggest that we develop an encoding of each tone into a sequence of tones, to avoid confusion. For example, we might encode tone A into A, but since the A tone may be received as a B, we cannot then encode B into the B tone, because then two encodings might be received as the same tone."

Smartee looked a little uncertainly at the admiral. "I understand, Smartee," Ecco whispered gently into his microphone. At that point, Smartee nodded his head with assurance.

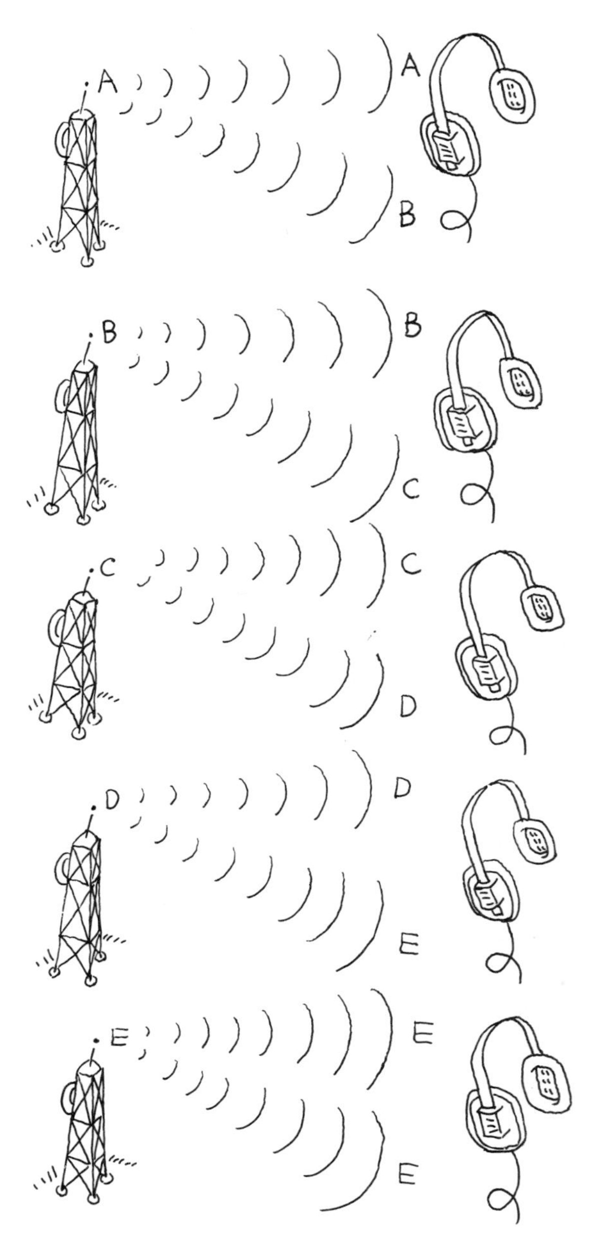

Symbols sent can be misinterpreted at destination. BCCAD was once received as CCDBE.

"Please think carefully about this problem, Dr. Smartee." pleaded the admiral. "It would be a tragedy if a war started because of a miscoding. Moreover, the code you discover should be as efficient as possible."

"I always think carefully, Admiral," responded Smartee coolly.

"Yes, of course," said the admiral. "I will return tomorrow."

? 1. Find an efficient encoding of the messages that leaves no possibility of confusion. How do you know there is no better encoding? Do not assume that pauses have any significance.

The next morning, Smartee gave Admiral Trober the answer that Ecco had arrived at after only a few minutes of thought. The admiral thanked Smartee for the solution.

"I have one more question," said the admiral. "Some of our older Trident submarines, because of their relatively unsophisticated receivers, may mistake the echo of a tone for two instances of that tone. It happens rarely and only at certain longitudes, but we don't want to make any mistakes. We will be replacing the electronics over the next few years, after which we will be able to use the encoding system you just gave me. But, in the meantime, can you design a code that will work well, in spite of tone echoes as well as tone confusions?

? 2. Try to find a minimal encoding for this situation. Again, your encoding should not assume that pauses are available.

Ecco told Smartee to ask the admiral to wait. Smartee did, and retired to the Idea Chamber. Within a few minutes, Ecco delivered a solution to Smartee. After a sip of cognac, Smartee emerged from the Idea Chamber and presented Ecco's solution to the admiral.

When we finished we celebrated this latest achievement over a delicious early lunch of fruit, eggs, and freshly baked bread.

15. The Territory Game

When a crisis arises, the omniheurist is in great demand. When nobody has a tough problem, the omniheurist must pursue other interests. We entered a period of quiet after the submarine problem, and for Evangeline other interests consisted of a systematic exploration of the grounds of Baskerhound's estate. Once I took a walk with her and she told me some of the things she had found out.

"Rumtopo gave this place to Baskerhound," she said. "That's one reason Baskerhound always invents games for him. The great appeal of El Casino to many tourists is the thrill of gambling in new ways."

We approached the main road in an area where the hedges were low. I saw a jeep drive by, and I immediately recognized the driver as Pedro Alcatraz, great-grandson of the pirate. He appeared not to recognize us.

"Do you think Alcatraz knows we're here?" I asked Evangeline.

"Yes, I'm quite sure he does," she said.

"Why hasn't he helped us?" I asked.

"The weeds brew," Evangeline answered. She did not amplify this cryptic remark, and I decided not to pursue it.

That evening Baskerhound brought a photograph of a new client to dinner. In his forties, Generalissimo Pedro Beluda of Argentina wore dark aviator sunglasses and a uniform with gold epaulets. Three rows of medals adorned his left breast, rewards for acts of bravery in a war — with the British over the Falkland Islands — that had ended in humiliating defeat. The generalissimo, however, distinguished himself by protecting the Argentine capital from an attack that never came. Now the generalissimo was the most powerful man in Argentina, leader of a group of generals who had seized the government in a bloody coup. It was said that Beluda regretted the ruthless brutality of some of his young officers, so he punished them by denying them their chauffeurs for six months.

Baskerhound explained the situation: The Argentine and British governments are negotiating the fate of 13 oil drilling sites near the Falklands — or the Malvinas, as the Argentines call them. The 13 sites are equally spaced along a line segment. The amount of oil under

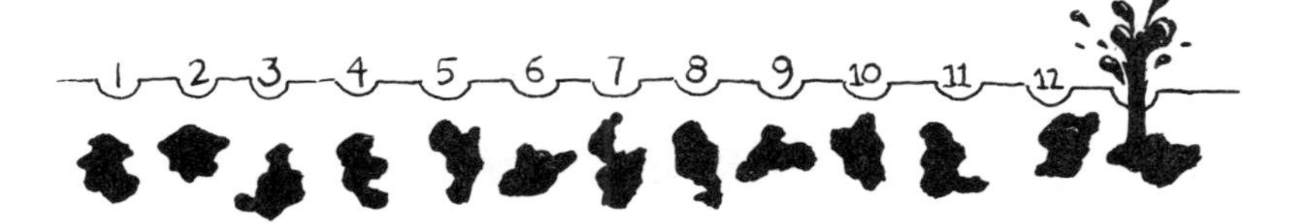

The 13 promising drilling sites are equally spaced along a line segment. There is about the same amount of oil under each site. What is the best way to play this game?

each site is thought to be roughly equal. The British have proposed a game as a way of dividing up the sites. The game will be played in rounds. In each round, the British lay claim to a hitherto unclaimed site, making it a B-site. Then the Argentines lay claim to a hitherto unclaimed site, making it an A-site. At the end, any B-site would belong to the British. Any A-site would belong to the Argentines. Any unclaimed site S would belong to the British if S is closer to a B-site than to any A-site. Similarly, an unclaimed site S would belong to the Argentines if S is closer to an A-site than to any B-site. If S is as far from the closest A-site as it is from the closest B-site, then neither side drills at that site. The two countries detest each other so much, they prefer no oil be drilled from a site rather than let the other side win it.

"The British have proposed this game and have asked to go first in each round. The good generalissimo wants me to help his country —and, I suspect, his own Swiss bank account—by determining the number of rounds Argentina should play to come out with at least as many sites as the British."

"Well, they will surely lose if they agree to only one round or to seven rounds," I volunteered. Evangeline glared at me. I shut up.

? Do you see why Argentina would lose if they agreed to one round or to seven rounds?

"By going in the center, of course" said Baskerhound. "But if the game were two rounds, then the British would certainly lose if they took the center site in the first round."

? 1. By playing for Argentina, how can you ensure that your side wins, if Britain plays first in the center in a two-round game?

"If the British were allowed to take two sites at once and then allow the Argentines two sites, the British would win," said Ecco.

? 2. How might Ecco show this?

"That is no help," said Baskerhound. "I am interested in alternating moves. Notice that the British can guarantee themselves a tie in a two-round game."

? 3. How can the British accomplish this?

"The real question," continued Baskerhound, "is whether the Argentines can guarantee a tie in two moves."

? 4. Can they?

Evangeline could take no more. "You are an unprincipled mercenary," she exclaimed. "You spout all this anarchic rhetoric, but you work for governments, for the military-industrial complex, and now for corrupt and vicious generalissimos—for anybody, provided you're paid well enough!"

"My dear Dr. Goode," said Baskerhound, "my actions are a means to an end. What a pity that I can't tell you what that end is. Maybe then you would approve of me."

"You do not understand the most basic lesson of history," Evangeline told him coldly. "Good excuses lead to bad ends!" She stood and left the table.

"What an empress she would make," said Baskerhound softly, and he watched her until she walked out of sight.

Contest Puzzle 3: Assume that there are 31 sites in the line instead of 13. Again, the British go first in each turn and then the

Argentines move. Can the British guarantee a win in a three-move game? Can they guarantee to win in a four-move game?

16. Drugs and Interdiction

We didn't see Baskerhound the next day. He must have arrived at some satisfactory conclusion with the generalissimo, because when he came home that evening he was wearing a medal. Evangeline glared at him. When we all sat down to dinner, Baskerhound removed the medal and threw it into the fire. "Agreed, Dr. Goode," he said, looking at Evangeline. "I will choose my clients more carefully in the future."

"Maybe there is some hope for you after all," Evangeline said in a conciliatory tone.

Baskerhound smiled happily. "That is cause for celebration. Shall we try our luck at El Casino?" he asked.

"Why not?" said Ecco, apparently pleased with the reconciliation.

El Casino had not changed much: the ladies wore the same expensive dresses, the men smoked the same expensive cigarettes. This time, however, Rumtopo greeted us personally.

"Welcome, Dr. Baskerhound, and welcome to your guests," he said with a smile that revealed his designer teeth. He pointed towards a corner of the room where many spectators surrounded a man who was standing next to a table, pressing buttons. "He is trying to control the coin-flipping machine," said Rumtopo. "As you can see, the Elves Flip game is extremely popular and has remained so, even after we reduced the odds to 6 to 1. Three-Finger Shooting has not yet caught on, but I expect it to be more popular when I reintroduce it as *Jeux des Trois Doigts* next season."

Rumtopo walked us to a corner table with an excellent view of the casino's activities and then left us. At the table next to us sat a short, fat man and two attractive women. Several bodyguards surrounded the party. We recognized him as the man who had asked Evangeline so many questions about the dice game Hi-Lo on our earlier visit to the Casino.

He stood and made a polite bow in our direction. "Dr. Goode, how delightful it is to see you again," he said. "My name is Solaris, though everyone calls me Jefito. How unfortunate that you must now accompany Baskerhound." He laughed.

"How did you recognize us before?" asked Evangeline.

"Simple," he replied. "There were rumors afloat that you had picked up Ecco's trail. So I knew to look for you and I knew what to look for." With that, he removed from his breast pocket a photograph of Evangeline and some other students taken at Oxford.

"How did you get this?" asked Evangeline, laughing.

"It appeared in the Rhodes annual report," he answered. "You see, I was also a Rhodes scholar once. I try to keep up."

He paused. He studied us. We studied him.

"You're wondering what I do now with my brilliant education, no doubt?" he said in answer to our unspoken question. "I am a cocaine broker."

Again he paused, smiling as he saw us stare in disbelief.

"It's a good business," he said. "In what other business would the police occasionally arrest my chief competitors? The police ensure my huge profits, and enforcement even helps create addicts. Since people are never sure when they'll get their next fix, they always want the most potent stuff we've got. They pay more and become loyal customers. Best of all, some of the cops actively help us. As one CIA *amigo* puts it, 'You have cash that cannot be traced. We give you protection. A simple quid pro quo.'"

Motioning us to join him, El Jefito sat down again with his beautiful assistants. The bodyguards remained standing, covering his back.

"Oh, yes," he said indicating the bodyguards, "it looks as if I lead a dangerous life. But guards are simply a cost of doing business. In truth, if one avoids excessive greed, one can live a long, healthy, and extremely satisfying life." He caressed the hair of one of his companions.

"Jefito," said Baskerhound, "you will be amused to know that the U.S. Air Force has asked Dr. Ecco here to help shoot down some drug-trafficking planes on their way to Texas."

"Oh, I don't use planes," Jefito said, smiling. "Suitcases and bribery are a much safer and cheaper alternative. So, please, be my guest, Señor Ecco and help Tío Sam. I could use more market share." With that, El Jefito rose and walked away, followed by his entourage.

The seven Air Force bases are evenly spaced along the northern bank of the Rio Grande. Each base controls the section of the border between it and the neighboring bases. How many planes are needed to catch the intruders?

"What's this about drug interdiction, Baskerhound?" asked Ecco, obviously annoyed. "I'm no pilot."

"Oh, the Air Force has many excellent pilots, my dear Ecco," said Baskerhound, "but they're short one mathematician. Here is the problem. They have set up seven air bases along the western Rio Grande. Each base controls that section of the border to which it is closest. A drug-trafficking plane (the Air Force word is 'intruder') flies into the section of one of the bases. The plane flies low and has a very small radar profile, so the central command knows which section the plane will pass only within eight minutes of its doing so. The intruders coordinate among themselves beforehand, but once they commit to a strategy, they don't change course for any reason. Eight minutes is just enough time for one or more jets to travel from a base's section to the neighboring base's section. A single intruder can elude a single jet, but it can't escape two jets. Two intruders can elude two jets, but not three. The Air Force wants no intruders to get through. Each night brings at most two intruders, so the Air Force's original strategy was to station three jets at each base. That way, each base could catch intruders without asking for help. But keeping twenty-one jets and crews in readiness in that region meant that other borders went unprotected. They would like to reduce the number to under ten if possible."

"I see how to do it with fourteen jets," I volunteered. "Keep two at each base. If a single intruder approaches, those jets can handle it. If a second should approach at the same time or later, a jet from a neighboring base can fly over."

"Nice observation," said Ecco. "Perhaps one can do better. Tell me, Baskerhound, you mentioned 'Central Command.' Does that mean that the Central Command gets all the radar feeds? Does it also mean they can recall a jet that's already on its way from a base to its neighbor?"

Baskerhound nodded.

"In that case," said Ecco, "I believe I have found the minimum possible. It is under ten."

He sketched out his solution for Baskerhound, who gave it to Kate Edwards to fax to Smartee.

? 1. Prove that this can be done with fewer than 10 jets.

? 2. Show that your answer is the best possible.

Jungle Killers

17. Amazon Exchange

Pragmatism is death.
Inscription on Baskerhound's Lear Jet

The week after the drug problem was uneventful and hot. I spent the mornings in the swimming pool and the afternoons in my room writing my journal and reading the U.S. newspapers in a vain attempt to determine whether the government had found the spy ship they had lost off Moher and whether anyone cared that we had been kidnapped. On Saturday afternoon, most of Baskerhound's thugs had gone to the nearby beach to chase after sunbathing Argentine beauties. The few remaining ones guarded the perimeter, holding their killer dogs at the end of short leashes and keeping an eye on Ecco. At 2 P.M., Evangeline walked into my room and said, "Professor Scarlet, let's take a walk."

"Now, in this heat?" I asked, looking at her over my newspaper.

She narrowed her eyes and nodded. I packed my journal, put on my sunglasses, and followed her outside.

We left through the front door and walked in a straight path until we entered the woods that sprang up about thirty meters from the house. Then we turned right and walked until we lost sight of the house, from which point we headed toward the back of the grounds. As we approached the edge of the property, I expected to hear a dog growl at any moment. But we made it through the hedge safely and found Señor Alcatraz—savior of bulls and Ph.D.'s—smiling at us from a jeep.

"I had to drug three of them and their dogs," Alcatraz told us. "But that wasn't hard. I've been giving the guards Belgian cookies by the box for the last few days. They shared them with their ever-hungry Dobermans. Today I sprinkled a few of the cookies with the hallucinatory powder distilled from the *Narcoplanta* weed. The guards should recover after a day or two of rest. Where is Ecco?"

"He was too heavily guarded to come," Evangeline replied. "He told me he would be all right. Baskerhound needs him too much."

Alcatraz took us to a large white house with a coat of arms crowning the entrance. "My house," he said when we arrived. "The bull-racing stadium that you have helped design is ten miles closer to Montevideo. What a pity that you cannot visit it. You must drive and quickly. Take the jeep. Brazil is due north. I don't recommend stopping until you are deep in the jungle—Baskerhound is sure to have you followed."

We heeded his advice. Four hours later we were sipping coffee in a café just outside Porto Allegre. A helicopter circled once low overhead and I could see Nigel Williams looking from its window, scanning the parked cars. Luckily, ours was under a tree. The helicopter continued to fly north.

"Gringos!" exclaimed a sunburned man, throwing up his hands in a gesture of evident contempt. "They fly to mines while we poor ones must walk through the jungle." I could see that his shirt had been patched many times.

"Mines?" I asked.

"Emeralds, my friend," he said. "The best emeralds in all the world." He pulled a ring off his callused finger and handed it to me. It consisted mostly of a large clear green stone which I took to be an emerald. I nodded knowingly, then gave it back to the man.

"From the jungle," he said pointing northwest. "But I'm not going back. Those wild men are killers."

"Who are they?" I asked.

"The Hagglitos," he replied. "They are more fierce than the mighty Yanomami. They don't even wait until you do something they don't like. They just attack."

I looked at Evangeline.

"We have no choice," she said, reading my thoughts. "If we go towards the coastal towns, Baskerhound will surely capture us."

I had to agree with her. We left our informant with money for an extra beer and drove our jeep towards the northwest. We took an unpaved jungle road a few hundred miles over the Brazilian border. The road crossed streams and muddy swamps. Often we would see animal carcasses half-buried in the mud, licked clean by the jungle and river predators.

The road ended at a village of thatched roofed huts with spears standing in front of them. As we entered the village, a man came running toward us. He looked very out of place, since he had sandy blond hair, wore a rain poncho, and carried a camera, a cassette recorder, and various pens.

"Who are you?" he demanded. Without waiting for an answer, he said breathlessly, "Thank God you're here. I'm Jack Raye. My collaborator, Carol Hale, and I are anthropologists from the University of Michigan. We arrived two weeks ago and are doing a linguistic and cultural study here among the Hagglitos."

A little calmer now, he told his story. "Last night, Carol disappeared. After much discussion with Chief Sassi Nosassi, I have come to understand that she is being held for ransom. The Hagglitos will exchange Carol for three gourds, a hairbrush, a Brazilian cruzado coin, and a machete. But I have none of these items. All is not lost, however, since I do have three bottles of aspirin. They never consume the stuff, but they believe it has magical powers and will put a few tablets in the thatching of their huts for its spiritual medicinal value. So, they'll exchange a hairbrush and a gourd for two bottles of aspirin. In fact, they have very elaborate and rigid rules of exchange. I have written down what can be exchanged for what on this sheet of paper."

As he pulled out a crumpled piece of paper from his left pocket, one of his aspirin bottles fell to the ground. Before he could pick it up, a village kid swooped it up and ran fast for the jungle.

"Two aspirin bottles, then," he said with resignation as he laid out the rules. "I've written down what can be exchanged for what, using two-headed arrows. I spent all night trying to figure out a sequence of exchanges, but no luck. Perhaps you can do it, whoever

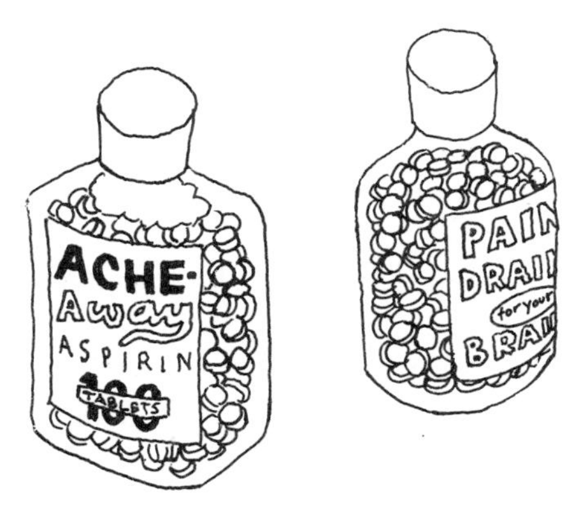

Does the Hagglito exchange system allow us to free Carol Hale if we have two bottles of aspirin at the start of bartering?

you are. The exchanges can go either way. A number in front of a letter means how many of that item must be part of the exchange. For example, and he pointed to the notation 2A↔N, "that means that two aspirin bottles can be exchanged for a spool of nylon fishing line. Each symbol stands for a particular item: A for aspirin, B for Brazilian cruzado, C for candle, D for dog, F for fishhook, G for gourd, H for hairbrush, M for machete, and N for nylon fishing line. The question is: Can you recover Carol starting with these two bottles of aspirin?"

This is what Raye had written:

Hale↔3G H B M

2A↔H G

A↔B C F

2A↔N

B↔M H

N H↔3C 4D

G↔2C 2D

2G↔N

M↔3N

M↔H 2C

A↔C

? Is it possible, by some sequence of trades, to recover Professor Hale? If so, show the sequence.

While the village children stared and giggled, Evangeline and I solved the problem. After Jack Raye executed the solution, Carol Hale was duly returned, completely unharmed. Chief Sassi Nosassi, a small but well-muscled man, came with her. He spoke to Raye in Hagglito, and Raye informed us that we had been invited to dinner.

Contest Puzzle 4: Starting with two aspirin bottles, can you rescue Professor Hale by some sequence of exchanges that uses six or fewer of the rules? If your answer is yes, then show such a sequence of exchanges, labeling each exchange with the rule that it comes from. If your answer is no, then show a sequence of exchanges that uses as few rules as possible.

18. Mutual Admiration

Dinner was delicious, grilled river fish garnished with lizard eggs. Through Raye, the chief asked us many questions. Were we tape-recorder people (the Hagglito term for visiting linguists and anthropologists) or were we prospectors? We told him that we were neither.

Our answer seemed to puzzle him a great deal. "What do you want from us, then?" he asked. "Gringos always want something."

"Just to stay with you a few days," Evangeline said.

The chief nodded but warned, "Beware of the assassins." With that, he rose. Raye and Hale also stood, we did the same. When the chief stands, the meal is over.

We followed the two anthropologists back to their tents. Raye's was the larger one, so we all agreed to meet in his jungle office. Carol Hale described what it was like to be held captive.

"They led me away yesterday afternoon, telling me that they had something to show me. I became completely lost as we walked through the jungle. At last we arrived at a small clearing where some men had begun to gather wood for a fire. We sat down and watched. When dusk came, they started the fire and I understood that we would not return that night.

"The next morning I woke up to find that the fire was still burning. They told me to walk around the fire as much as I could. As I circled the fire ever more slowly, they ran around it, singing prayers to the gods of the forest. Finally, I sat down and refused to go on. Some women appeared and began to anoint me with perfume. It suddenly occurred to me the fire and the perfume might be connected. I started to panic. I shook off the women and began asking for the chief. They told me he would come soon. Just as I was contemplating the likelihood of my surviving a dash into the jungle, a little boy came running through the forest with three gourds, a hairbrush, and a machete. The men around the fire cheered and began to douse the fire. The women put me on their shoulders and carried me back into the village. So thanks to you, I'm alive."

"We're glad to have been of help," said Evangeline. "But we are a little worried about this talk of assassins. Could you tell us what the chief meant by his warning?"

"Well, first, there really are assassins here," answered Raye. "Our colleague Steve Miller died at their hands. Assassination is a cultural ritual for the Hagglitos. By a selection process that we don't fully understand, certain males become assassins and others do not. The killings occur when there is a clear full moon. The next full moon is due in two weeks.

"One thing that Miller discovered was that the nonassassins—the 'good people' in Miller's terminology—will tell you the truth about good people and about bad people, or assassins. Bad people may lie. He gave the name of 'mutual admiration clique' to a group in which every individual says that everyone else in the group is good. He also found that there are some good people—nonassassins—in this village, though the chief may not be among them."

Carol Hale took up the story. "All assassins are males who have passed the rite of manhood. That rite, in case you are curious, is to capture a python and bring it back to the village alive. Anyway, there are 25 such men in the village. We have interviewed all of them. If we can discover who the nonassassins are, then we can survive by inviting them to dinner on the night of the full moon."

"How will that help?" I asked.

"According to Hagglito tradition," she replied, "if a host invites only nonassassins to dinner on the night of a full moon, then the host is invulnerable to assassination attempts. The belief is so deeply held that assassins will not even dare to attack for fear of avenging spirits."

"Have you interviewed the men to find out what claims they make?" I asked.

"Yes, we have interviewed all 25 of them," Raye responded. "Here are our results. The chief is number 1 and the newest initiate is number 25. We must know who is definitely good and who is definitely bad:

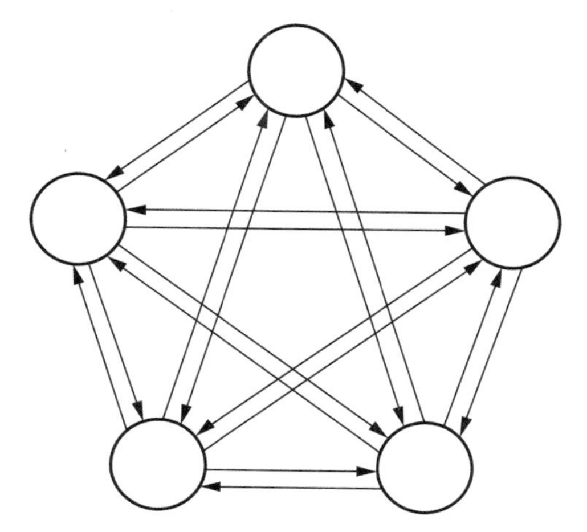

Each arrow represents the assertion, "That person is good." Here is a mutual admiration clique of five Hagglitos. In the village there are 25 possible assassins, but at least one of those 25 is not an assassin. Which one or ones?

1–5 constitutes a mutual admiration clique, but they claim that 18 is bad and 14 is good.

6–8 are also members of a mutual admiration clique, but they claim that 13 is bad and 3 is good.

9 and 10 form a mutual admiration clique.

11–20 is almost a mutual admiration clique. The exception is that 18 and 19 accuse one another of being bad. Further, 15 asserts that 9 is bad.

21–25 form a mutual admiration clique, and 22 claims that 9 is good and 11 is bad."

? Who is good for sure? Who is bad for sure?

Evangeline solved the problem within the hour. The anthropologists were happy to learn that they had at least two good people to invite. They had extra sleeping bags, so Evangeline slept in Carol Hale's tent and I slept in Jack Raye's.

The next day I asked Evangeline what she thought we ought to do. "I've already done something," she said. "I attached my pip-card to the roof of the tent."

"Your what?" I asked.

"The transmitter," she answered. "Recall the card that the Director gave me before we set off for Uruguay the first time. He claimed that it had a range of two hundred miles. I've tried it before. It didn't bring us help in Thailand and it didn't bring us help in Uruguay, but maybe this time it will."

We spent the next five days following the anthropologists as they studied kinship relationships and marital taboos. As the days passed, I grew more and more worried. Suppose Miller was wrong, and there were *no* good people in the village. He had obviously been wrong about something.

But the fifth day brought a welcome, if noisy, surprise. A Brazilian army helicopter thundered overhead in the late morning and landed in the middle of the village. The officer in charge approached the four of us. "Professors Goode and Scarlet?" he asked. "I am Lieutenant Panta Leao. The Director greets you."

He invited us back to the helicopter. We thanked the anthropologists and asked them to join us. They declined. "We will survive," said Carol Hale. "This culture will not. Soon they'll be selling their machetes to tourists carrying credit cards."

Double Escape

19. The Octopelago Problem

Small problems can take
more effort to solve than big ones.
From Ecco's notebook, "The Duel"

Lieutenant Leao's helicopter took us to the airport at Rio, where we found tickets for the nightly Varig plane to Washington waiting for us. Never had lukewarm, overcooked airplane food tasted so good. After two months spent facing Baskerhound's goons and the jungle's terrors, I was eating food from a plastic tray and breathing filtered air. To many people this would not be an improvement over raw nature, but to me it brought a rush of nostalgia. I would have welcomed my electric bill at that moment.

As we left the plane in Washington, two middle-ranking navy officers greeted us. The elder of the two handled the introductions: "Professor Scarlet, Dr. Goode, my name is Nicholas Chase, captain, U.S. Navy. This is Commander Victor Shaw. The Director sends his regards." They then escorted us in silence to a stretch limousine whose passenger compartment consisted of two leather couches facing one another.

As we drove away from the airport, Chase spoke. "We are taking you to the Washington Meridien Hotel, where the Director will meet you and debrief you over breakfast. However, we must ask your help during the drive to the city. You see, we are in charge of the project to recover the U.S.S. *Freedom.* We now agree with you that Baskerhound may be connected with its disappearance. Recently, there has been some suspicious activity in Micronesia that suggests the presence of a hidden ship, possibly the *Freedom.* We would like to send Commander Shaw to take a closer look. He needs a cover that will enable him to travel all through the islands, so we have represented that he is an expert at solving transportation problems. We happen to know that the local government has been trying to solve a problem with ferry service among the islands. If Commander Shaw can become involved and solve that problem, that will give credence to his cover."

"Captain, I don't suppose this can wait until we've had a night's sleep," I said, forcing myself to stay awake.

"I'm afraid not, Professor," said Chase. "Commander Shaw is scheduled to leave this afternoon."

The captain explained the problem. "The ferries serve a part of the Truk district of Micronesia known as the Octopelago, a group of eight islands regularly spaced as in vertices of an octagon. In 1964, the governor decided that it was too risky to send canoes from one island to another. So a ferry service began between every neighboring pair of islands on the octagon. Each of their eight ferry boats went back and forth between the same two islands day after day, charging one dollar for each one-way trip. There were two problems with this service. First, collecting the fares every trip was very burdensome for the collector. In fact, the collectors went on strike, demanding that the government implement a one-way fare scheme in which no fare would be collected in one direction, but twice the fare would be collected in the other. Their slogan was 'One way or no way.'

"At first, the government protested that they might lose money in the bargain. For example, if they only collected fares in the clockwise direction, they argued, then a frugal islander might take ferries only in a counterclockwise direction. However, a young island mathematician designed a one-way fare scheme such that any trip by ferry that began and ended on the same island would cost the same as if the previous two-way fare scheme were in effect."

? 1. Can you think of such a one-way scheme?

Captain Chase continued. "The second problem was that the trip was very long; each ride took at least an hour. The schedules were irregular, so getting to a far island took a long time."

"Why didn't they just have all eight ferries go back and forth to some fixed pickup point?" I asked, hoping he hadn't already answered the question.

"Well," Chase said, "there were two problems with that. First, most passengers only wanted to go to a neighboring island. Your design would lengthen that trip. Second, and more important, the ferries were quite old and an open-sea journey might cause them to sink. Recently the government bought two new ferries. These new boats are quite a bit faster in the open sea, though near the islands they are as slow as the old ferries. The new ferries can make the trip between any two of the islands in an hour. The government wants to make sure that no trip requires more than three ferry rides. Equally important, they want to maintain their one-way collection scheme, *and* maintain the equality between the one-way scheme and a two-way scheme. That is, any trip by ferry that begins and ends on the same island should cost the same as a two-way fare scheme in which each ferry ride costs a dollar.

In which direction must the fares be collected so all passengers pay what they should in any round-trip journey? What should be the routes of the two new ferries?

2. Can you design routes for the new ferries (leave the old ferry routes alone) such that no trip requires more than three ferry rides? Also, your design should include a one-way fare scheme that is equal to the two-way scheme in the sense stated.

"We appreciate your help very much, Dr. Goode," said Captain Chase after Evangeline had solved the problem. "Now, we must ask you one more favor. The islanders also have a quite different problem concerning air service. If Shaw does not learn enough about the *Freedom's* disappearance by pretending to study the ferries, he will stay on to solve the air service problem. I've enclosed the problem description in this envelope. A courier will pick up the solution from the hotel concierge at 1000 hours tomorrow. Oh, I've also included a sizable cash payment. We do appreciate your help."

He handed the envelope to Evangeline—it was crammed with hundred-dollar bills. At that moment, the limousine arrived at the hotel and a doorman opened the car door. As we stepped out, a man in a dark blue jacket with a white shirt and straight striped tie nodded to us in greeting. "My name is Steven Underton, Special Agent FBI," he said. "Please follow me. The Director is waiting."

20. MicroAir

The Director had no official government position, but everyone in the Washington establishment knew of him. The use of the unspecified title "the Director" conveyed a sense of his importance in a town where the cognoscenti spoke of the Agency (the National Security Agency), the Company (the Central Intelligence Agency), and the Bureau (the Federal Bureau of Investigation). He now sat at the breakfast table in the hotel restaurant, dressed in a freshly pressed suit and solid red tie. In front of him, we saw the morning's *Washington Post* and a closed folder labeled "Secret."

He wasted no time. "What can you tell us?" he asked.

"Smartee is a charlatan," I said immediately. "Ecco, Baskerhound, and occasionally Evangeline and I have solved all the difficult problems."

"I'm not surprised," said the Director. "He disappeared about three days ago. We found elaborate communications equipment in his so-called Idea Chamber. What else?"

We described our escape. The Director seemed surprised we had managed so easily. "Must be distracted," he mumbled.

Then we told him about Koh Samui and Punta Ballena while Special Agent Underton took notes. The Director promised to send commandos to both sites. "I don't think we'll find anything though," he said. "If Baskerhound suspects that you've survived your escape, he will know to avoid his habitual haunts. By the way, did Baskerhound ever mention Micronesia?"

"No," said Evangeline. "He ranted a lot, but never told us anything truly secret. When he transported us, he just packed us up into his LearJets and off we went."

"LearJets," said the Director. "That's interesting. Yesterday two LearJets crashed within sight of one of our cruisers in the Gulf of Thailand near Cambodia. That's a strange place for that type of plane to fly, especially in the current political climate there. The cruiser found no survivors, but wasn't able to approach the coast too closely." He whispered something to Underton, who nodded and then left the dining room. "Well," said the Director, "the question now is what to do with the two of you. You may be in danger. Baskerhound knows we are after him. I can offer you protection."

"No, thank you," said Evangeline. "We would prefer to continue our quest to liberate Ecco as unobtrusively as possible."

The Director frowned. "Dr. Goode," he said, "your courage is admirable, but if you plan to leave the country to find your friend . . . "

"Rest assured, sir," said Evangeline. "We will stay in the country."

"As you wish," said the director. He jotted something on a piece of paper and handed it to Evangeline. "Here is a new number. Call in case you're in trouble."

After we finished our breakfast, in silence, the Director tossed me a key. "It's to a suite," he said. "You deserve a little luxury after what you've been through. By the way, your help with Shaw's cover is much appreciated."

We went up to our room and I slept soundly for the next eighteen hours. When I awoke, I found Evangeline already working on the second problem that the Navy men had given us. Their memo read:

TO: Dr. Evangeline Goode and Professor Justin Scarlet
SUBJECT: Micronesia Airlines

Micronesia Airlines serves the seven principal islands in Micronesia. The airline, small but proud of its efficiency wants to guarantee that a trip from one island to another will take no more than two hours and will not require changing planes. Further, they want to ensure that there is a flight from any island to any other every three hours. So, a passenger should be guaranteed to be able to get to any other island within five hours of the time he or she arrives at the airport without having to change planes.

It takes about one hour to fly from one island to another, so there should be at most one stop in any flight. Remember that no passenger should have to change planes. The airline has just purchased three additional planes and now has seven.

? 1. Can they schedule the plane to guarantee that these conditions are satisfied?

? 2. If one plane is unavailable, design routes to guarantee that every passenger can travel from any island to any other island within 5½ hours of arriving at the airport possibly by adding stops to some trips, though still permitting no transfers?

We left the solution in a sealed envelope with the concierge before 9 A.M. and took a taxi to the train station. Evangeline bought two unreserved seats and we boarded the 10 A.M. train to New York. Evangeline insisted we take seats at the rear of one of the cars.

Within fifteen minutes of our departure, she asked me to walk with her to the café car. When we returned to our seats, she asked me whether I minded moving to another part of the car. A few minutes after we sat down, she suggested a second trip to the café car. I said I wasn't thirsty. She said, "Let's see whether they sell playing cards." Again, we took new seats when we returned.

The seven principal islands of Micronesia have airports. If MicroAir has seven airplanes, how can flights be scheduled so that a passenger is guaranteed to arrive at his destination island within five hours of arriving at the starting airport and without changing planes?

"We're being followed," she told me in a whisper.

As we approached Philadelphia, she suggested yet another trip to the café car. This time I didn't object. We were still in line when the train arrived at the station and continued to stand while the train was unloading. As soon as the last new passengers boarded, Evangeline tugged at my sleeve and ran off the train onto the platform. I ran after her. We looked up and down the platform, but no one had jumped out after us. I followed her up the steps to the street and into the back seat of a waiting automobile.

21. Personals

The driver was a gaunt Chinese man in his early thirties. I studied him in the mirror as he drove swiftly and expertly through the downtown traffic. He had serious, intelligent eyes and a long thin scar starting at the bridge of his nose and slanting down to just under his right cheekbone. Evangeline kissed him gently on the back of the head when we stopped briefly for a light. They exchanged a few words in Chinese.

"Evangeline, what is going on?" I asked. "What are we running away from?"

"Sorry, Professor," she said. "Here is the plan. We will stay with my cousin Zhongli here and keep moving from house to house—Zhongli has many friends. We'll take a roundabout route to New York, and call the *Times* to leave a message for Ecco in code in the personals column to tell him we are nearby."

"But why?" I demanded with some irritation. "Why not return to our apartments and accept government protection? Ecco will know where to find us."

"Professor," replied Evangeline calmly, "things are not as simple as they seem. Try to imagine how Baskerhound's mind works. If we accept government protection, that means that we believe Baskerhound to be dangerous and that we are ready to cooperate fully with the authorities to catch him. But remember that Baskerhound believes himself to be an idealist. He would most likely consider such behavior on our part to be treachery. Ecco might suffer the consequences."

"Unlikely but plausible," I admitted. "Now, please introduce me to your cousin."

"My apologies," replied Evangeline. "Zhongli Chang, this is Professor Justin Scarlet."

Chang turned his head slightly and nodded, but kept his eyes on the road as he continued to weave skillfully through the traffic. "I have taken the American name of Michael, if you find that easier to pronounce than Zhongli, Professor," he said in only slightly accented English.

"You see his scar?" Evangeline asked me quietly. "Notice that it begins on the bridge of his nose and slants off to the left, making him look like a street fighter. He did indeed get it in the street—the streets of Beijing. He was caught by soldiers of the People's Liberation Army as he carried a wounded friend to the field hospital during the Tiananmen Square protests. One soldier, with a fixed bayonet came towards Zhongli—Michael. Michael backed away. The soldier sliced at him for sport, breaking the bridge of his nose and slashing his cheek. Then he laughed and kicked my cousin in the ribs. Fortunately, the soldiers were called away by their commander, so both Michael and his friend survived."

Michael interrupted, "No pity, please, Professor. I was one of the lucky ones. If I had gone to a hospital on the day of the massacre, my name would have been kept on a special list of suspicious people. Such people are never given exit visas. Avoiding the hospital let me emigrate. It was worth a badly healed nose."

We drove out of Philadelphia and into New Jersey. Evangeline wrote out a message for Ecco: "We are on the corner of Monmouth and Main."

Michael drove us to an apartment house at that location and settled us in to an apartment he'd borrowed from a friend. "Professor," said Evangeline, "I wonder if you could help with my encoding effort. Here is how the code works. It begins with the encoding that Ecco began with in his first letter to us. Then, each time a letter is used its rotation encoding is increased by one. So, if the message were 'call a cab' it would be encoded as 'cwhi x dyd.' The first 'a' becomes a 'w,' the second becomes an 'x,' and the third becomes a 'y.'"

? 1. What is the encoding of the message Evangeline wrote to Ecco?

"How will Ecco respond?" I asked after handing the encoding to Evangeline.

"He will use the same technique, except that he will start with the encoding that we finished with," Evangeline said. She verified that my encoding agreed with hers and then called the *Times* to place the message. She motioned to me to pick up an extension.

"I'm sorry, ma'am," said the personals editor, "but your message is not in English. We don't even know what language it is in."

"You're right," said Evangeline. "It's a code between me and a foreign spy."

The editor chuckled. "And I suppose the foreign spy will respond in a similar vein?" she asked.

"I hope so," said Evangeline.

"Very well," said the editor, "it will appear in tomorrow's edition."

We bought the *Times* the next day at an out-of-town newspaper stand, but saw nothing of interest besides our own message. We moved every twenty-four hours in the middle of the night, driven by Michael in a zigzag course to New York. On the eighth day of our wandering, at a stand in Summit, New Jersey, we picked up the *Times* and saw the following:

lf zqauzmoga pn kaobg. ppq hfb qghh. rdei.

? 2. Decode this one. Remember that its encoding starts where Evangeline's left off.

22. Shark Labyrinth

After solving the puzzle, Evangeline and I thanked Michael and took the next train to New York City. Michael drove off to the south.

We took a taxi to the Village. The drive brought a new wave of nostalgia over me. Yes, the city was dirty and many people looked like thugs in black leather jackets with protruding spikes, torn jeans. But the city was alive. Signs advertised new plays, new dance performances, experimental music. What I liked best wasn't advertised: performance art and chess. The taxi took us down the west side of Washington Square Park and I looked at the crowded chess corner. My favorite adversary, the "captain," who had a scraggly beard and wore ten rings on each hand was not there, but other regulars were—like the tall Serb who played for $40.00 a game and kept his Great Dane Apollo at his side.

The taxi dropped us in front of Ecco's MacDougal Street apartment. We climbed the stairs to his third-floor apartment and knocked. A few seconds later, a smiling Ecco, ginger cookie in hand, opened the door.

"How good to see you back home," Evangeline said happily. "We're free at last."

"Or at least for the moment," said Ecco, the ever-cautious mathematician.

"How did you escape?" I asked.

"Please sit down," Ecco said. "It's a long story."

We sat around his dining room table and joined him in cookies.

"Baskerhound was furious with Williams when he learned of your escape." Ecco told us. "Williams tried to blame the guards, but Baskerhound would have none of it. Baskerhound took me and the most trusted members of his staff back to Thailand and told Williams to follow in four days. During the first half of the flight Baskerhound just sat brooding. We didn't even eat together. About midway across the Pacific, he picked up the airplane's phone and dialed a number. He asked, 'Are you ready?', then paused for a moment. Then he said, 'Good. Be off to D.G.' He looked much happier after this brief conversation and fell asleep. By the time we were back on Koh Samui, he was completely at ease, as if a great weight had been lifted from his mind. During lunch, he patted me heartily on the back and said, 'Ecco, we've shared some amusing puzzles, haven't we? If, for some reason, we don't see each other again, please bear me no harsh feelings. I will be away for a few days. You will be lightly guarded until Williams arrives, but please don't try anything crazy. You know about the sharks.'"

Ecco stoked up with another ginger snap and continued. "He left that afternoon. Without Baskerhound to keep him in check, Williams might have tried to exact vengeance on me for your unscheduled departure. So, all I could think of was to escape. Kate Edwards invited me to her office for tea that afternoon. Just as I sat down, she heard the whistle and rushed out. I looked among her bookshelves and soon discovered a map entitled 'Labyrinth of the Sharks.' This described the routes and resting places of the sharks that Baskerhound maintained beyond the barrier. I hid the map underneath my windbreaker. When Kate returned, she told me that Smartee had left Washington to fly to Geneva. She said that he had heard of your escape and was afraid that you would denounce him as a fraud. 'Would serve him right, too,' she added.

"After returning to my hut, I studied the map. It consisted of a network of corridors along which the sharks could travel. The sides of the corridors went from sea floor to sea surface. The network was made up of equilateral triangles. To my great surprise, I noticed a door at the intersection of two corridors—a sea exit. The network, I decided, provided both protection and an escape route. I went to the kitchen, where I knew there was a small collection of spear guns for fishing lying around. I took two and concealed them in the woods. Then I borrowed a scuba-diving outfit from Kate, telling her that I planned a dive among the coral the next day. I hid that in the woods as well. So far, so good. But there remained the problem of the sharks.

"I remember that Baskerhound had told me that his pet sharks were great whites, formally known as *Carcharodon carcharis.* I looked them up in a reference book that I found in Baskerhound's library. As adults, these sharks are six meters long, weigh two tons, and possess serrated triangular teeth that are four times as long as human teeth. They are called great whites, because their gray-blue back blends into a snow white belly. They can attack in less than three meters of water. They are killing machines before they are born. When a female gives birth, the first baby sharks to leave the womb have often devoured their siblings while still inside. Sharks have good eyesight, in spite of what people believe. They steer the final fifteen meters to their prey using their eyes and electrical sensors in their snouts. Beyond that, they use hearing and, above all, smell. Fortunately, if you should find yourself in the water around Koh Samui, they are unlikely to smell you more than thirty meters away because the water is quite still. They are not picky eaters. A great white was once caught with a keg of nails and a few bottles of beer in its stomach. Except in a feeding frenzy, they are quite cautious about attacking humans. But in a frenzy, the sharks lose all inhibition. They bite victims and one another at random. A shark in a frenzy has been known to bite its own tail, gorging on its own flesh."

Evangeline shuddered. I felt a bit queasy myself.

"That evening I had dinner with Kate Edwards," Ecco continued. "'I'm missing the shark map,' she said with a smile as she had finished her meal of dolphin fish steak. 'If you are planning a swim, you might need this.' She slid a waterproof diver's pack across the table. I

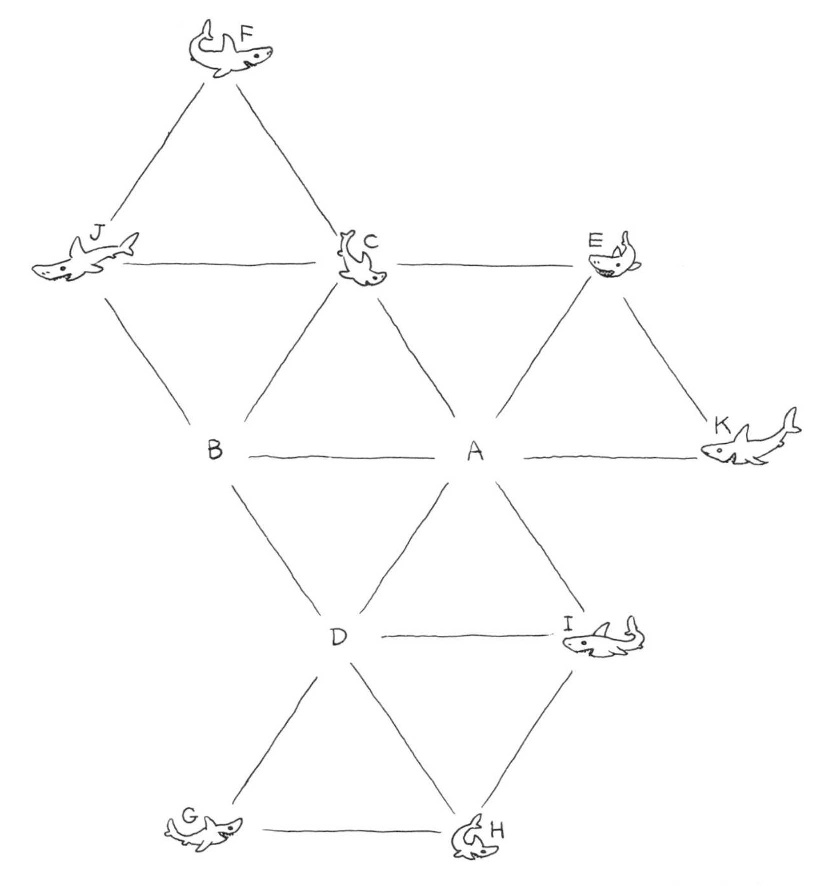

The map of the shark corridors; each fish marks the home of a shark. Ecco must enter at A and leave at F. All edges are 100 meters long.

started to thank her but she shook her head abruptly, stood up, and walked back to her hut. I wrapped several pieces of leftover dolphin fish into a napkin with ice and took them back to my hut. When I opened the diver's pack, I saw it contained a few thousand dollars in Thai and U.S. Currency. I awoke the next morning at 6 A.M. and walked into the woods to hide. My plan was to be out of the labyrinth by the time the 8 A.M. launch headed for the mainland. I studied the map once more. Here, I saved it to show you. All edges of the network are 100 meters long. I've sketched the sharks' positions. I was to start at A. The exit was at F. Based on what I had learned about sharks, I formulated the following rules:

1. I will be attacked if I'm within five meters of a shark, but am safe if beyond that distance.
2. Under no circumstances should I dare to swim past a shark that is swimming in the opposite direction in the same corridor.
3. A spear-gun can kill a shark within 10 meters. I have two spears.
4. The dolphin fish will draw a shark 90 meters from its lair, unless there is other blood in the area.
5. Sharks can swim 600 meters in the narrow passageways of Baskerhound's underwater habitat in the time it takes me to swim 100 meters.
6. I can swim 200 meters in the time it takes for a group of sharks to finish devouring a wounded shark or the dolphin fish.
7. A shark will not leave its lair for blood that is more than 200 meters away.
8. Once at the exit, I am safe because the exit has a door.

? How did Ecco get through the labyrinth?

"Everything went according to plan. I arrived at the launch on time and caught the flight to New York three hours later."

"Bravo, Ecco," I said, clapping hands. Evangeline kissed him gently on the forehead.

"Now tell me about your adventures," said Ecco.

We told him about our escape, our stay with the Hagglitos, our rescue thanks to the pip-card, our meeting with the Director, and our encounters with Michael. We had just finished our story, when the phone rang. Ecco turned on the speaker-phone. It was the Director.

"Ecco, welcome back. Please excuse my late call, but passport control is so slow these days in providing us with the most basic information. Your friends gave us the slip, but I assume they are with you by now. Please congratulate them on their prowess in solving the ferry and airplane puzzles. It seems that the governor begged Commander Shaw to stay behind to solve a myriad of other problems that the islands faced. But I didn't call to exchange pleasantries," he continued, unaware of any irony. "I have good news. Our cruiser off Cambodia picked up wreckage from a LearJet this morning. It bore the insignia of Baskerhound's flagship plane. Good riddance to that troublemaker, eh?"

He hung up. We sat in silence for more than a minute. I think each of us, even Evangeline, had felt some affection for our captor. Could he really be dead?

23. A Question of Inheritance

The next day, news of Ecco's return was in all the papers. Journalists besieged his apartment, but, true to form, he refused to grant interviews. In the weeks that followed, Smartee's would-be clients, reading that it was Ecco who had solved the problems credited to Smartee, begged Ecco to come to their aid, proferring huge sums of money to entice him. Ecco turned away nearly all such clients with the excuse that he needed rest. But he didn't rest. Instead, he collected and studied books and magazines about military, particularly naval, hardware. One of the books was entitled *Diego Garcia: Strategic Linchpin of the Indian Ocean.* Dry stuff, I thought, very dry.

There were some cases Ecco couldn't refuse.

The telegram arrived at Ecco's MacDougal Street apartment in the late morning. Postmarked Dublin, it read: "A dispute over a will. Please take tonight's Aer Lingus flight. Will meet you at arrival gate in Dublin. O'Getman."

"Are you free?" Ecco asked me. I nodded. Evangeline had been asked to give a plenary talk in the upcoming Logic in Phenomenology conference, so she couldn't join us.

In Dublin, the inspector looked fit and cheery in his new chief inspector's uniform. He led us to his car and began his tale.

"The Earl of Braycot," he explained, "earned the affection of all the people of Dublin. He was generous, witty, and a symbol of hope for many. You see, he was paralyzed from the waist down, yet never complained and always found ingenious ways to overcome his disability. Many of his inventions, such as the famous Braycot Arm-Walker, have been patented all over the world. His children, on the other hand, are an irresponsible lot. While they were at Trinity, it was a rare month when some Braycot didn't turn a seventeenth-century dormitory room into a Donnybrook Fair. The earl died of heart failure Wednesday last. It was the night of his seventieth birthday, so at least he went out in style. From the looks of it, he may well have the last laugh. You see, his will challenges his children to a puzzle, a variant of which he had once proposed in an international contest. I understand, gentlemen, that the two of you won that contest."

The only difficult puzzle in the contest had been the one submitted by the earl, which I called the Fakes Puzzle:

> You are given 20 coins. Some are fake and some are real. If a coin is real, it weighs between 11 and 11.1 grams. If it is fake, it weighs between 10.6 and 10.7 grams. You are allowed 15 weighings on a scale (not a balance). You are to determine which of the 20 coins are real and which are fake.

"As you will see," said the inspector, "The earl has proposed a variant of his puzzle in his will."

Our car had begun to climb a hill. At the top stood a castle surrounded by a wall and a moat. We drove across the drawbridge, parked in the courtyard, and entered through the main gate into a paneled foyer. Paintings of Braycots through the ages gazed at us from four sides. O'Getman led us to the sitting room which had a large round table at its center. The will was spread out on the table.

We read:

Dear Children,

This puzzle will separate the boasting philanderers from those who are made of studier stuff . . .

Then the puzzle was stated, but instead of twenty coins there were only 10, and the contestant was told that there were at most two fakes among the 10.

Now, heirs, each of you must state the minimum number of weighings sufficient to find the answer. Those who conclude the correct number will divide the inheritance. The others will get nothing. All of Irish society will see which Braycot really has a head on his shoulders or Lady Luck on his side. In life, the two are equally important.

When we had finished reading, the inspector said, chuckling, "Well, there are going to be some poor Braycots in this part of the world. The earl's ten children guessed every number between three and ten, with two of them guessing seven and two guessing six. I say guessing, but I might as well say shouting. Each of them delivered a spirited polemic stating why his or her answer is correct, but none of them has a proof. Finally they agreed to ask you to be the judge. Your fee will enable you to take a few years off."

Ecco nodded in gratitude, then asked, "So you want me to demonstrate that some number of weighings is enough no matter which zero, one, or two coins is fake. Then you want me to show that fewer weighings are not enough, is that right?"

Of 10 coins, 2 at most are fakes. What is the smallest number of weighings sufficient to find the fakes?

"Yes, sir, and quickly," responded the inspector, listening to the loud voices in the room next door. "The young Braycots are in a wild mood."

"Well," Ecco said, turning to me, "the rules haven't changed, so we can weigh no more than three coins at once to be sure of the number of fake and real coins in a weighing."

"Yes, that's right," I said. "For example, four coins weighing 10.7 grams will weigh the same as three coins weighing 10.6 grams and one weighing 11 grams.

"So let us see how well we can do by weighing just three coins at a time," Ecco suggested.

? 1. Find the smallest number of weighings under the assumption that at most three can be weighed at once.

As I began to work on that problem, Ecco thoughtfully devoured a few biscuits. Then he began making scribbles of his own. An hour later, I found a solution and presented it to him.

"Well done, Professor," he said. "However, true omniheurists must answer the question in full generality. What is the smallest number of weighings that are enough without making the three-coin assumption?"

? 2. How many weighings are sufficient then?

24. The Toxicologist's Puzzle

The next day, the would-be heirs of the Earl of Braycot heard the news in uncharacteristic silence. Some beseeched Ecco to reconsider his answer, presenting one foolish reason after another why their own answers were more worthy. Ecco surprised me with his patience, but finally asked Inspector O'Getman to drive them away.

The three of us then celebrated with the inspector at his favorite pub, Mary Ann's. After two drinks Ecco fell asleep. Inspector O'Getman and I had a riotous evening, but I don't remember leaving the bar. In fact, I don't remember anything but hearing innumerable folk songs sung drunkenly but in key and being shaken vigorously in my hotel bed the next morning.

"Stop pounding on me," I am supposed to have said. I do remember feeling as if I were being hit with a beer mug.

"Wake up, Scarlet," Ecco said, continuing his shaking. "I thought you could hold your liquor."

I sat up and also saw Inspector O'Getman with a twinkle in his eye and a pleasant smile. He must have drunk twice as much as I, but to look at him one would have thought that he had bedded down at 8 P.M. after saying his prayers. Next to him stood a somber looking fellow.

"Our services are needed, Scarlet," Ecco said. "It seems that the earl may not have died a natural death."

The third man spoke up. "I'm Raymond Dooling, the chief forensic toxicologist of Dublin. I don't know how much you gentlemen know about toxicology," he said in a morbid tone.

I looked at him blankly.

"Many people think that looking at corpses is unpleasant work, but it is vital, gentlemen, very vital. It is necessary to be able to detect poisons, if for no other reason than to bring murderers to justice. Recall how toxicology halted the rash of undetected alkaloid murders in the early 1800s."

I was in no mood for a history lesson, but as usual Ecco was curious. "Please tell us what happened," he said.

"As late as 1847," our visitor explained, "the world's forensic toxicologists had no way of coping with the new chemical poisons derived from exotic plants. The major ones known were morphine, strychnine, quinine, and nicotine. The chemists called them alkaloids. No method was known to isolate them from corpses in the bodies of the murdered, so deeply did they penetrate into the tissue. It took a Belgian of special talent, Jean Servais Stas, to isolate nicotine by a method that has relevance to our problem today. Luck played a large role in his success. The body of the victim, one Gustave Fougnies, had been drenched in vinegar by the murderers to mask the smell left by the nicotine. The police followed normal procedure by storing the body in alcohol. The combination of vinegar and alcohol proved fortunate. The alcohol-acid mixture penetrated the body and dissolved the alkaline vegetable poisons. The alcohol could then be left to evaporate and the residue washed with water. Only the poison dissolved in water. Potash then made the alkaloid poison precipitate from the water."

Chemistry had never been my strong subject. I barely restrained myself from asking, "And who was Potash?" My look of puzzlement did not stop Dr. Dooling's droning, however.

"Well, both poisons and our methods have improved, Dr. Ecco," he said. "But the spirit is the same. The earl's body contains traces of a poison that is unknown, but we know its possible constituents, which appear to be variants of arsenic, bismuth nitrate, cadmium chloride, diproprionitrile, and ethyl arsenate. Let's call them by their first letters, a convenient set: A, B, C, D, and E. We need a testing strategy to determine which of these constituents is present in the body. We know now that at least one is present.

"One complicating factor is that two other compounds may also be present, one containing BC and the other containing ABC. Both compounds are formed spontaneously and rapidly in liquid solution. If there are fewer A molecules than B molecules and fewer B molecules than C molecules, then ABC will form until there are no more A molecules, and the remaining B molecules will form a compound with some of the remaining C molecules.

"Our tool is a set of five membranes that act as filters. Filter 1 will let any of B, C, or D pass through, but is impermeable to A or E and to any compound. Because the compounds form spontaneously, what passes will not have *both* B and C molecules. We denote this by saying filter 1 is permeable to {B, C, D}. Filter 2 will let {A, B, C, BC} through, but nothing else. Filter 3 will let {B} through, but nothing else. Filter 4 will let {B, D, E} through, but nothing else. Filter 5 will let {A, D} through, but nothing else."

Ecco interrupted, "Let's do a trial run. If the only possible constituents were A, B, and D, for example, then would the following strategy work?"

? Try to devise a strategy for that case.

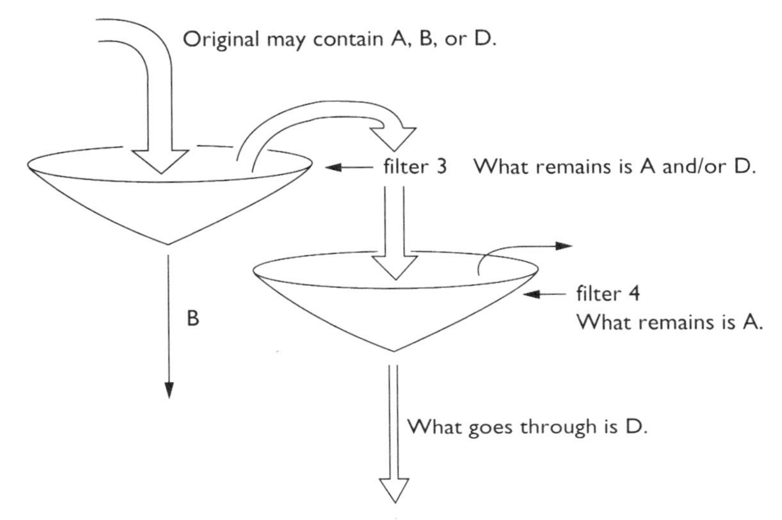

Ecco's proposed solution to the trial-run problem.

Ecco outlined his plan. "Put the solution through filter 3. If anything goes through, it is B. If there is anything that does not go through, then A and D are both possible. Put the remainder in filter 4. If anything goes through, it is D. If anything does not go through, it is A. So, we can determine which of any of {A, B, D} is present, if they are the only possibilities."

"Exactly," said Dr. Dooling. "The trouble is that the membranes work very slowly. Each takes nearly a week to allow the compounds to penetrate. Also, they are very expensive and our supply is limited to one of each type except that we have two of filter 5. Can you determine, Dr. Ecco, which constituents are present within three weeks using the available membranes?"

? 1. Can you do it?

Ecco immediately began working on the problem and soon arrived at a solution. Dr. Dooling thanked him and apologized that thanks were all he could offer as the city was far more frugal when it came to paying consulting omniheurists than were putative heirs of the Earl of Braycot.

"It is always a pleasure to struggle with such fine puzzles whether they bring material rewards or not," said Ecco graciously. "Should you have any more, please let me know."

His wish was soon granted. We were packing our bags and Inspector O'Getman waited to take us to the airport when Dr. Dooling came to the door. He handed the inspector a note. After reading it, O'Getman chuckled. "Eire is loathe to see you go, my good Dr. Ecco," he said.

"Some of the Braycots are bent on leaving the country," Dooling explained, "and we must let them go in one week if we have no evidence. On the other hand, the spectroscopy laboratory concluded that neither ABC nor BC is in the body. With the filters we have to offer you, can you find the constituents in a week?"

"I'm afraid not," said Ecco immediately. "One week is too little time, even without those compounds."

? 2. How does he know?

Dr. Dooling was disappointed until O'Getman recalled that Article III, Section 2.5 of the Irish Criminal Law permitted the chief city detective to prevent suspects from leaving the country for two weeks in the event of chemically induced deaths. O'Getman said he thought he could persuade the chief detective.

While Ecco was thinking about the problem, I realized that if spectroscopy could limit the search to just one constituent, then three filters and one week would be enough.

? 3. Which three filters?

I never did find out what Ecco's conclusion was, but it must have satisfied our hosts, because Ecco's name was all over the newspapers the next day. "The Great Omniheurist Solves Again," pronounced *The Dubliner.*

? 4. Do you see how to do it?

Ever camera-shy, Ecco grumbled about the publicity as we made our way to the airport. Fame, however, has its advantages. Upon presenting our tickets to the agent, we realized that we had forgotten something. "Gentlemen, your flight was yesterday. These tickets are no longer good." he said. "I must check with the supervisor."

"Please allow me to speak to him," said Ecco. "I think I can explain everything."

No explanations were necessary. The supervisor had just finished reading the morning newspapers. "Please come this way, sir. It is an honor to have you fly with us," he said as he readjusted the tickets, changed them to first class, and returned them to Ecco, all within a minute. "There you are, sir," he said, "quick and sweet like an ass's gallop."

Battle for a Continent

25. How to Steal a Submarine

What is history but
a fable agreed upon?
Napoleon Bonaparte

Midway on our flight to New York, the captain made an announcement: "Ladies and gentlemen, due to unusually heavy air traffic in New York, we're being rerouted to Washington, D.C. We'll land and refuel before taking off again for the Big Apple."

I groaned, but Ecco leaned back and looked dreamily at the illuminated seat-belt symbol. "Curious," he said. "Our plane should have enough reserve fuel to circle New York if necessary."

Just before we landed, a stewardess handed Ecco a note. "We have just received this radio message for you, sir," she said.

Ecco showed it to me: "Ecco, Baskerhound has reappeared. You're needed. Disembark at Washington. We will be waiting for you." There was no signature.

The two officers who had met Evangeline and me after our rescue from the jungle greeted us at the gate. "Captain Nicholas Chase, Commander Victor Shaw, this is Dr. Ecco," I said by way of introduction.

The officers escorted us in silence to the customary stretch limousine. In the car, Captain Chase explained the situation. "On his mission to the islands, Commander Shaw did indeed find signs that the U.S.S. *Freedom* had been in Micronesia: an abandoned dock with modern parts in the Marshalls, a large unpaid fuel bill to some unknown English sailors, and a few other indications," he said. "But the ship had already left. Apparently she was disguised as a fishing vessel. She must have changed her disguise at least once more. A few days ago, the U.S.S. *Groton*, one of our old Trident II submarines, mistook her for a submarine tender, a ship that repairs submarines at sea. The *Groton* surfaced and the crew of the *Freedom* boarded and took control of the submarine."

"You mean Baskerhound, or his agents, now controls a ballistic nuclear submarine?" I exclaimed.

"That's right, Professor," said the captain. "They even know how to launch the missiles. Intelligence says the submarine has only a skeleton crew, but they only need five people to fire the missiles."

"Where and when did they capture the submarine?" Ecco asked.

"Two days ago, near our naval base on Diego Garcia in the Indian Ocean," answered the captain.

I glanced at Ecco. He smiled slightly. Had he known this was going to happen? How much more does he know? I wondered.

The limousine whisked us to the Pentagon, where Admiral Trober and the Director were waiting for us. Admiral Trober put the problem bluntly: "Somehow, Baskerhound fooled a submarine into thinking that his little spy ship was a new kind of submarine tender. Dr. Ecco, we need your help to determine how he did it. We know this much: he must have used our newest 'authentication protocol.'"

"What is that?" Ecco asked.

"An authentication protocol consists of a proof that some agent is who he claims to be. All of our protocols work like this: if A wants to communicate with B, A sends a message that all ships in a certain area may hear, but that only B will understand. B may respond with a signal directed only at A—so no other ship will hear it—called a *private signal.* After a few more messages, known as challenges and responses, each ship is convinced that the other is not an imposter, or, if you wish, that the other ship is authentic. Except for the first broadcast transmission, all transmissions use a private signal. We thought that we had a protocol that would prevent a would-be imposter from masquerading as another vessel V either by listening in to some broadcast transmission involving V or by participating in some exchange with V.

"The basic scheme that we use is called public key encryption. We deemed this the most secure, because, even if our enemies capture one ship, they will not know the private key of any other ships. Here is how public key encryption works. For each agent A, there is a secret key S_A that only A knows and a public key P_A that all ships know about. The book of those public keys was on the spy ship that Baskerhound managed to capture near the Cliffs of Moher, along with the private key for that spy ship. Public and private keys combine in the following way. If m is a *cleartext* message, that is, one that anyone can understand, then applying either key to a message m yields an unintelligible message. We represent the application of S_A to message m by $S_A(m)$. Similarly, $P_A(m)$ represents the application of P_A to m. However, if S_A is applied to $P_A(m)$, yielding $S_A(P_A(m))$, then the result is simply m. Similarly, $P_A(S_A(m))=m$. We have arranged the codes so that it is impossible to derive S_A, given P_A. It is even impossible to figure out S_A given $S_A(m)$ and m."

The Admiral continued, "The first protocol we developed allowed A to broadcast to B the message $S_A(P_B(\text{I am A}))$. Depending on the strength of the transmission, many ships may hear the broadcast. But only A would be able to make that statement and only B would be able to decode it by applying P_A and then S_B. The idea was that B would then send a private signal directed at A."

"One of my mathematicians quickly found a flaw," said the Director arrogantly.

? Can you see any flaws?

Unruffled, the admiral went on: "An imposter, C, could masquerade as B by simply responding to A's message. Our second protocol solved this problem by having B respond $S_B(P_A(\text{I am B}))$."

"We cracked that one, too," chuckled the Director.

Basic authentication protocol:
1. A broadcasts, "I am A and am looking for B."
2. B sends privately, "I am B, now prove you are A."
3. A sends privately, "Here is proof that I am A."
Did the Navy's version of this protocol give too much away in the proof?

? Can you see how?

"I think I see how," said Ecco. "An imposter, C, could simply remember the first message that A sent, for later use. When C broadcasts that message later, agent B would then believe that it was responding to A, when it was in fact responding to C."

The admiral, still unflappable, said, "That's right. Our next idea was to use *nonces.*" He saw the puzzled looks on our faces and was about to explain. The Director interrupted rudely.

"And I though you were an educated man, Ecco. A nonce is a large random number that is used only once by any single agent. Because it also encodes the ship identification number, no two ships ever use the same nonce either. Understand?"

Ecco exchanged glances with the admiral, who flushed red for a moment. Then he continued: "At this point our protocol became, with n being a nonce,

i. A sends to B via broadcast: $S_A(P_B(\text{I am A}))$;
ii. B responds to A (privately): $S_B(n, P_A(\text{I am B}))$;
iii. A sends to B (privately): $P_B(S_A(n))$.

Our best belief is that this last protocol is secure, because the imposter has no way of duplicating the third message. Can you find any holes in it?"

I saw the Director smile.

Ecco thought for a moment, then asked for some cookies. When an aide brought some and he had taken a bite, he said, "Even if I accept your assumptions that the codes can't be cracked, I see a way for imposters to gain the trust of other ships, provided the ships send messages to one another fairly frequently."

? 1. How would that work?

"That must be how Baskerhound succeeded," said the admiral.

My mind wandered back to Baskerhound's theater in Punta Ballena and the magic tricks. Is this what Baskerhound meant when he said to Ecco, "You have changed the course of history?"

"Tell me, Dr. Ecco," the admiral said, "assuming we see our way through this crisis, is there any successful protocol you can devise? We have a new tool at our disposal. We have a global network of satellites that send clock signals every millisecond. All ships receive each signal as fast as light travels."

"What is the shortest time for a ship to receive a signal and then send the same signal to other ships?" Ecco asked.

"For technical reasons that I can't reveal," answered the admiral, "such a rebroadcast takes at least a second."

"Then it's easy," said Ecco.

? 2. In that case, can you figure out a scheme that is cheap and imposter-proof? Why does the one-second delay matter?

The admiral seemed satisfied with this solution. "We will begin using that protocol starting tomorrow," he said. "We certainly don't want to lose another submarine this way."

On our way to the airport to resume our interrupted flight home to New York, the Director lectured us. "Dr. Ecco and Professor Scarlet, what you heard today is extremely sensitive," he said. "Only a handful of people in the military and the executive office know about it. The public will become uncontrollable if it learns it is threatened by an errant submarine, in spite of the emergency measures that we have begun to put in place. Revealing what you have heard to anyone, Dr. Goode included, is treason. Understand?"

Similar threats must have kept everyone quiet, because no news of the submarine kidnapping appeared in the newspaper for the next four days. On the fifth day, Baskerhound arrived in New York and called a press conference.

26. Missile Roulette

"Ladies and gentlemen of the press," said Baskerhound. "I am in control of a nuclear submarine, formerly called the *Groton,* now known as the *Anarchist.* It is a Trident II submarine, and we have reprogrammed its missiles to aim at a few selected cities, towns, and farms all over the world. We will begin firing these missiles unless our demand is met."

Within seconds, the major networks interrupted their regular programming to broadcast this announcement. Within minutes, the police, under government orders, moved to stop the press conference, but it was too late.

"What is the demand, Dr. Baskerhound?" asked one reporter.

"We want the governments of the world to cede Antarctica to the new state we have formed. That state is called Free."

"The entire continent?" asked the same reporter.

"All of it," answered Baskerhound serenely. "No scientific stations, no fishing stations, no more exploitation of the krill, nothing. Free will be a new kind of state, dedicated to a pure environment and individual freedom. People from anywhere on earth will be invited to join us on Free. They should be of sound mind and body, demonstrate a respect for nature, and believe in liberty."

"By when must your demand be met, Dr. Baskerhound?" asked another reporter.

"Unless I extend the deadline, in five days," said Baskerhound.

"And if your demand is not met?" said the reporter.

"Every day, I plan to send a one-character code via ELF," Baskerhound answered, "the extremely low frequency transmitter that the United States uses to communicate with its submarines. Every day the character changes. There is no pattern—the character sequence was randomly generated. If the *Anarchist* does not receive the proper character, it will begin to fire its missiles. Of course, I have memorized the sequence."

The police broke up the press conference and Baskerhound was escorted by FBI agents to a waiting plane. Thirty minutes later, President Nebraska Yerrek announced to the nation: "My fellow Americans: We have begun negotiations with Dr. Baskerhound and are confident that we can resolve this crisis peacefully. Dr. Baskerhound has requested wide access to the media during these negotiations and we have agreed to that request. For the duration of the crisis, Congress has enacted certain emergency laws. Please keep to your normal routine as much as possible. Thank you."

That evening, Beverly Waters had a special interview with Baskerhound. "Dr. Baskerhound," she said, "people say that you are an idealist in spite of your terrorist threat to us. Are you?"

"Ms. Waters," Baskerhound answered, "history tells us that political systems arise by force of arms. Consider the American War of Independence, the French Revolution, and the dictatorships that have plagued this century. To found an environmentalist anarchic state, I too had to resort to force of arms. Call me what you like."

"What led to your belief in anarchy?" she asked.

"Again, we must look to history," said Baskerhound. "Until this century, the mass of people did not resist oppression because they had someone else over whom to exercise authority: men over women, mothers over children, children over other children. It was a caste society in all but name. But castes made sense then: when the technology was primitive, most jobs were unpleasant. Caste systems assigned to each person a set of jobs. Since people had no choice, they felt little resentment. But modern technology frees us from the necessity of the tyranny of caste. Unpleasant jobs are fast disappearing. Yet governments perpetuate the lines of authority left over from the caste system. To me, anarchy is the only possible response."

"Dr. Baskerhound," the interviewer asked, "you call your country Free and seem to expect applications from people from all over the world. But how do you expect to attract them? Not many people have a burning desire to live in Antarctica."

"We don't want many people, only a few," said Baskerhound. "Free will be a center of learning and research. It will provide moral guidance to the entire planet as we travel through the next millennium. We think there are enough idealists who might wish to join us."

"Dr. Baskerhound, you are a philosopher by training," said Waters. "We know something about your political and environmental philosophy, but what is your philosophy of action?"

Baskerhound responded, "B. F. Skinner, despite his professional contempt for free will, once said, 'If an idea possesses you, drop everything and pursue it.' Not bad for a behaviorist, but it is too pessimistic for my taste. My philosophy of action is more direct: follow your dream until you achieve it."

The interview was translated into ninety languages. The Protectors of the Truth, a fundamentalist group, labeled him the Antichrist. Another fundamentalist group, the Children of God, considered him the herald of the Second Coming. Greenlove, the militant environmentalist group, declared, "Finally, environmentalism has muscle." Other environmentalists praised his goals but criticized his methods. Political commentators generally agreed with *Le Monde's* editorial accusing Baskerhound of "threatening to destroy the world in order to save it."

Negotiations among the powers that controlled and claimed Antarctica quickly degenerated into a shouting match. Many countries demanded compensation from the United States for the bases and stations they were to lose, arguing that since the *Groton* was an American submarine, the United States was to blame. The United States refused, protesting that such bargaining would waste precious time.

Two days later, Ecco received a call from Beverly Waters. "Dr. Ecco," she said, "Baskerhound will be on the show again tonight and has requested that you appear with him. He said something about extending his deadline if you win at a game."

Waters's show that night was called "Missile Roulette." The interviewer opened the show with a question. "Dr. Baskerhound, you have refused to name the targets of your missiles. Why?"

"Suppose you are in a room with a thousand people," said Baskerhound, "and you hear: 'Someone in this room will be killed.' You shudder with fear. After all, it could be you. On the other hand, if you hear, 'Someone whose last name begins with a V will be killed,' you breathe a sigh of relief. Perhaps you even convince yourself that the victim deserves it. Similarly, if I said that I have aimed the missiles at certain cities or certain countries, everyone else would shortly simply wish them good riddance."

"Sir, you asked Dr. Ecco to join you this evening," said Waters. "Why?"

"The world governments have two more days to meet my demands," Baskerhound answered, "but they have reached an impasse and have asked for an extension. I will give them an extension if Dr. Ecco wins a simple little game of chance and skill. Here is how it is played. I have brought thirteen cards with me, the ace through king of hearts. I will lay them down on this table in an organization that only I know. Ecco's task will be to guess which card is the ace. If he succeeds, I will extend the ultimatum. If he fails, I won't."

He laid out the cards face down. The camera moved in on them. As experts later verified, they were unmarked.

"Now, Dr. Ecco," said Baskerhound. "Here are the details of the game. You point at one of the cards. Maybe it is the ace and maybe it is not. I will turn eleven of the other cards face up. None of them will be the ace. So, two cards will remain face down—the one you pointed at initially and one other. You may choose either one of them. If you choose the ace, you win. There is one small catch. If you win by

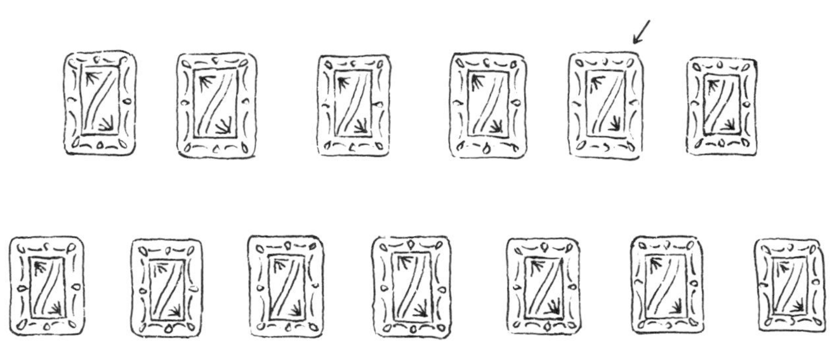

Initially all cards are face down. Ecco points to the card indicated by the arrow.

Then Baskerhound turns up all cards except the one Ecco pointed to initially and one other. None of the face-up cards is the ace. Are the two face-down cards now equally likely to be the ace?

choosing the one you pointed at initially, then we play the game only once. If you win by choosing the other face-down card, then we play the game twice and you have to win both times."

Ecco shook his head. "I'm confused," he said. "Are you saying that the first time we play the game, if I win by choosing the card I first pointed at, then you will extend the ultimatum? If I win by choosing the other face-down card, then I must win a second game for you to extend the ultimatum? If I lose in any game, then we stop playing and the ultimatum is not extended?"

"Right," said Baskerhound.

? 1. How should Ecco play to maximize the likelihood of extending the ultimatum?

Ecco played and won.

The headlines proclaimed, "World Praises Great Omniheurist" and "Risky Strategy Pays Off," and even "Does Lady Luck Smile on Omniheurists?" Ecco chuckled when I showed him the newspapers.

"Luck played only a small part," he said. "My strategy gave me overwhelming odds of winning."

? 2. What were his odds?

27. Finding the Target

Baskerhound happily assumed the role of what Tom Foxe called a "post-modern, apocalyptic cultural icon." He refused interviews that lasted more than five minutes and insisted on direct quotes. One of them was, "Absolute power corrupts absolutely, but I can take it," his quip to *Vanity Fair.*

In the meantime, the world powers continued to bicker. The United States had already begun withdrawing some of its scientists from Antarctica, as had the Soviet Union. However, much remained to be done.

With a week left until Baskerhound's second deadline, Captain Chase came to Ecco's apartment. "Dr. Ecco, Professor Scarlet, and Dr. Goode, please come with me," he said. A limousine took us to a nearby military airfield and a military jet flew us to Washington. Captain Chase briefed us once we were in the air.

"I'm not sure how much you know about antisubmarine warfare," he began.

"I've been reading up lately," said Ecco.

"Well, then you know that one approach we use is to overfly an area with a P-3C Orion airplane equipped with a magnetic anomaly detector," said the captain. "The detector has a range of a few hundred meters to the right and to the left of the plane and can distinguish the magnetic signature of a submarine from natural phenomena in the ocean. The beauty of this form of detection is that it is completely passive. The submarine doesn't even know the plane is there. By sheer luck, an Orion found the *Groton* last night and we have been tracking it ever since. Admiral Trober will give you the exact status of our tracking effort. I suggest that you rest now, because you have several difficult hours ahead of you."

But we barely had time to doze, since our jet completed the flight in less than forty minutes. A limousine with police escort was waiting at the military airport. We sped to the White House.

Admiral Trober, the Director, and a pale President Yerrek were waiting for us. Admiral Trober was the first to speak. "Welcome, Dr. Goode, gentlemen," he said. "You know that we found the *Groton* last night. Since then it has proceeded to the sea floor near Madagascar and has apparently stopped. Using passive sonobuoys, Orions, and 688-class attack submarines, we have certain information about it. An elite company of Navy Seals is in the area, ready to dive to the submarine and disable its missiles. They must work quickly however and for them to do so, we must figure out exactly where the submarine is.

"We know that there is an old sailing wreck in the area at location X. Passive sonobuoys tell us that the *Groton* is 9000 meters from X, but don't tell us the direction. The ocean floor in the area is almost

The Groton *is 9000 meters from the wreck at X. But in which direction?*

perfectly flat except for some underwater boulders that we have labeled B through F. Here is what else we know:

Boulder D is 5000 meters from X;

D is 3000 meters west of B;

D is 1000 meters from F;

E is closer to F than X is;

E is 4000 meters from B;

E is 5000 meters from D;

B is 4000 meters from X;

E is 1000 meters from the *Groton;*

B is 3000 meters from C;

C is 5000 meters from X;

F is a little to the north but mainly to the west of C and is about 6000 meters away.

"Can you tell us where the *Groton* is to within a 200-by-200-meter area?"

? Where is the Groton relative to X?

Evangeline and Ecco began discussing the problem over paper and pencil. After one false start, they gave an exact fix on the *Groton.* Realizing the consequences of a mistake, all three of us rechecked their reasoning several times.

When we were satisfied that their result was correct, Admiral Trober forwarded the conclusion to the waiting commandos. We waited tensely for two hours, worried that the submariners might fire their missiles.

After two hours, the radio receiver came alive. "Mission accomplished," cracked the voice. "The submarine is militarily useless. We even disabled the torpedos."

Everyone in the room cheered. Seconds later, the Director left. President Yerrek shook hands with all concerned, approaching Ecco and Evangeline last. The president no longer looked pale. "You may have saved millions of lives," he told them. "On behalf of all those lives saved, thank you."

The president turned to his chief of staff. "Arrest Baskerhound," he said. "Report to me when you have him." Then he left the room.

Uneasy Peace

28. Epidemiologists

Sire, it is not a revolt,
it is a revolution.
Duc de Liaucourt to Louis XVI,
July 1789

Back in New York, we saw that like most newspapers, the *Post* devoted a banner headline to the news: "GROTON IS NEUTRALIZED; BASKERHOUND SLIPS NET."

What should have been a tidy ending had come undone. Even though the president had ordered him arrested, Baskerhound was still at large. The newspapers all told the same story. During the crisis, Baskerhound had been staying in Smartee's Georgetown mansion. As long as the *Groton* had been armed and poised to strike, the government accommodated Baskerhound's every desire; not to do so might have led him to unleash the submarine's missiles.

But the FBI kept Smartee's house under continuous surveillance. FBI agents had glimpsed Baskerhound intermittently the whole day of the attack on the *Groton.* They watched him give press conferences, walk in the house, and speak on the telephone. The last time they saw him was ten minutes before the Navy Seals had finished their work. But when the FBI agents converged on the mansion within minutes of the disarming of the *Groton,* Baskerhound was nowhere to be found.

"No effort will be spared in locating Dr. Baskerhound," said an FBI spokesman.

Few people were worrying about Baskerhound's whereabouts. The world was jubilant that the nuclear threat had been eliminated. The Navy enjoyed most of the credit for the neutralization of the *Groton* and the capture of its crew, although President Yerrek told the press, "The people of the earth owe a great debt to Jacob Ecco and his colleagues. Omniheurism made the operation possible, first by extending the ultimatum and second by helping to locate the *Groton.*"

Adulation for Ecco poured forth from all over the world. True to form, he refused to give interviews or accept medals. He asked that any money awards be given to the biochemical study of learning.

His omniheuristic practice boomed. No longer did his clients come only from the desperate and incautious. Now appeals for help came from countless governments as well as from IBM, Mitsubishi, and Daimler-Benz. Unlike Smartee, Ecco refused cases that did not interest him, but he still had an enormous case load.

The first case he took came from the Centers for Disease Control (CDC) in Atlanta. Scientists at Superdrug, Inc. had invented a vaccine for the toocie virus, but it was expensive to produce. The CDC needed to understand better the transmission of the virus in order to decide how much serum to order.

Dr. Robert Yaman explained the situation to Ecco, Evangeline, and me: "You know that toocies are extremely contagious. Casual contact, even a handshake, with a diseased person or a carrier is enough and most people are susceptible. On the other hand, a few people are immune. Knowing who they are may help us find a cheaper vaccine. These are the four types of candidates:

1. The Sick, who have the disease, are contagious, and suffer symptoms—are now magenta.
2. The Carriers, who have the disease and are contagious but suffer no symptoms.
3. The Immune, who neither have the disease nor are contagious, even though they may have been exposed to it.
4. The Unexposed, who neither have the disease nor are contagious, but may be susceptible.

"We've had a small outbreak in Philadelphia. We are studying it to determine who had toocies first and who is immune. Here's what happened:

Mary met Keith on Saturday.

Bob met Alice on Saturday.

Leah met Ellen on Saturday.

Bob met Mary on Sunday.

Mary met Ted later on Sunday.

Alice met Keith on Sunday.

Who had toocies first? Who is immune?

Keith met Leah on Monday.

Bob met Leah later on Monday.

Still later on Monday, Ellen met Ted.

Leah met Mary on Tuesday.

Later on Tuesday, Leah met Bob.

Leah met Alice on Wednesday.

Ted met Alice on Thursday.

"At the end, everybody but Bob and Keith has the disease, though some are not yet sick. As for Bob and Keith, one is immune and the other is susceptible—we don't know which because somehow the hospital mixed their blood samples. Assuming that one person had the disease to begin with, who was it? Also, can you help us determine who is immune?

? Try to answer Dr. Yaman's questions.

29. MarsRail

Many commentators asserted that the Baskerhound threat had brought the people of the world together. "We have survived," the Asahi (Japan) *Daily News* editorialized, "Now, we must face our destiny as a species. The space station on Mars is the first step."

Innumerable other articles hailed the multinational Mars Station effort that would put a permanent community on the Red Planet. The effort presented some nasty technical problems, so we were not surprised when Aerospatiale, the French aerospace giant, requested Ecco's help, and off we went.

The launch facility on the Côte d'Azur was a wonderful sculpture of steel and hoses. Robots were everywhere, welding and carrying parts of cryogenic magnets. Dr. Patrick Robin got right to business. "*Docteur* Ecco, you know that the Europeans are to design the ground transportation for the Mars Station, no?"

Ecco smiled sheepishly and shook his head. "I hadn't even heard of the Mars Station before receiving your letter."

"*Aucun problème,*" he said. "Sorry, no problem. I fill you in on the details, yes?"

Ecco nodded.

"There are eighty and one huts we will place on the plains surrounding Olympus Mons," Robin said. "All the visitors ask why there are so many. We plan nine communities, and each has nine huts.

Some are small factories, others are dormitories. Others are stations of science. You understand? Astronauts will travel from one hut to another by the rail system we are designing. The trains will be levitated using superconducting magnets. Because tremors and meteors are a major danger, we want neither bridges over tracks nor do we want trains to change tracks. Instead, each track will connect two huts, no more no less. Four tracks can be connected to a hut. Astronauts will voyage from one hut to another, change trains, then continue on their way."

A crane load passed near our heads. Instinctively, I backed into the shelter of a nearby building, just as a truck came speeding out. "I am sorry very much," said Robin. "Our—how you say—deadline approaches, and we are very pressed."

Ecco seemed unperturbed. "Let me see if I understand," he said. "You have 81 huts. You connect pairs of them by a single track. The tracks lie on the plain. No bridges or tunnels are allowed. Is that correct?"

"*Exactement,*" said Robin. "Changing trains takes a long time, so it must be possible to go from any hut to any other through at most five intermediate huts. That is, you use six trains. Of the same spirit as your structural design for General Scott's Antarctic Research Station, *n'est-ce pas?*"

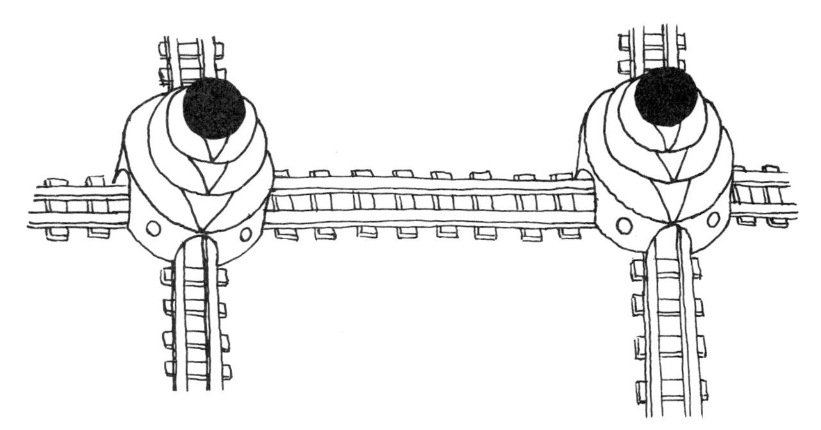

Each hut can handle up to four tracks. The tracks may curve, but no two tracks may cross. It should be possible to travel from one hut to any other by going through at most five intermediate huts.

"That's true," said Ecco. "Except that you offer me tracks instead of doorways, so the huts need not touch and the tracks need not be straight. Am I free to lay out the huts in any way I please?"

"But of course," said Robin. "We will give you complete authority if you can just show us a design with 81 huts satisfying these constraints. Our engineers haven't been able to accommodate more than 53 huts under these constraints. But the magnificent Dr. Ecco can perhaps do better, no?"

Just then a robot passed us and sideswiped Dr. Robin, knocking him down. "*Merde,*" he said as he picked himself up. "They have no manners, those steel-hearts."

? Can you find an 81-hut design to satisfy these constraints? Recapitulating, they are: maximum of four tracks per hut, it must be possible to go from any hut to any other using at most six trains, no bridges over the tracks. (If you happen to know graph theoretic terminology, try to construct an 81-node planar graph having diameter six and degree four.)

Ecco produced such a design after several hours' work. When Robin saw it, he embraced Ecco warmly in the French style and said, "*Magnifique, c'est vraiment formidable.*"

30. The Prince's Problem

On the day we were to leave France, Ecco received a delivery by courier. In the pouch he found a letter and two first-class airplane tickets to Emiremir, the richest principality of the Arab Emirates. The letter was from Prince Q'at Alhoun, a man the whole world knew.

His full title was Prince Q'at Alhoun, Sovereign of the Mighty and Protector of the Weak. His oil-rich principality granted its citizens free education, health care, and housing. With the substantial money left over from this generosity, the prince enjoyed commissioning palaces, a hobby for which he spared no expense, as we were soon to find out.

On a first-class flight to Emiremir, each passenger reclines on a leather chaise. A water pipe stands in easy reach for the pleasure of smokers, a side table holds dates and grilled lamb. Hand-knotted Tabriz rugs cover the floor of the cabin from wall to wall. Flight attendants serve fruit drinks and apricot-rosewater pastries. We arrived at our destination content and rested.

A man in an embroidered robe covering a western suit greeted us. "Dr. Ecco, Professor Scarlet, my name is Amman Bra'man. I am a cousin of the prince," he said. "Please follow me." He led us to a waiting limousine and we drove swiftly to the Prince's unfinished summer palace. Construction had stopped for the day, but the palace's graceful arches were already in place. Many huge rectangular tiles were stacked in the center of a large room.

We were led to a carpeted office where the prince awaited us. He nodded his head in greeting and we did the same. "I have commissioned a square interior court in my new palace to be 16 meters by 16 meters," said the prince after tea. "I have commissioned inlaid marble tiles, each 1 by 4 meters, to be crafted by the famous Venetian artisan Italo Perfecto. Perhaps you saw them just now. As you know, we Arabs love algebra and geometry, and I have given much thought to the possible tile patterns. But my plans have not pleased Allah, for Italo Perfecto died after completing only 63 tiles. His method died with him. I have prayed and thought about what to do with the 63 tiles and have decided that the design lacks the dynamic flow of water. I have therefore commissioned local masons to build four fountains. Each fountain uses a square meter of floor space. However, my great court mathematician hasn't been able to find a design that covers the courtroom floor with 63 1-by-4-meter tiles and places fountains in the four corners.

"I ask you now either to find a design with the fountains in the corners or to prove to me that none is possible. If none can be found, please find a design that uses all the tiles and that has fountains in as many corners as possible."

? 1. Is there a design using 63 1-by-4-meter tiles with one fountain in each of the four corners?

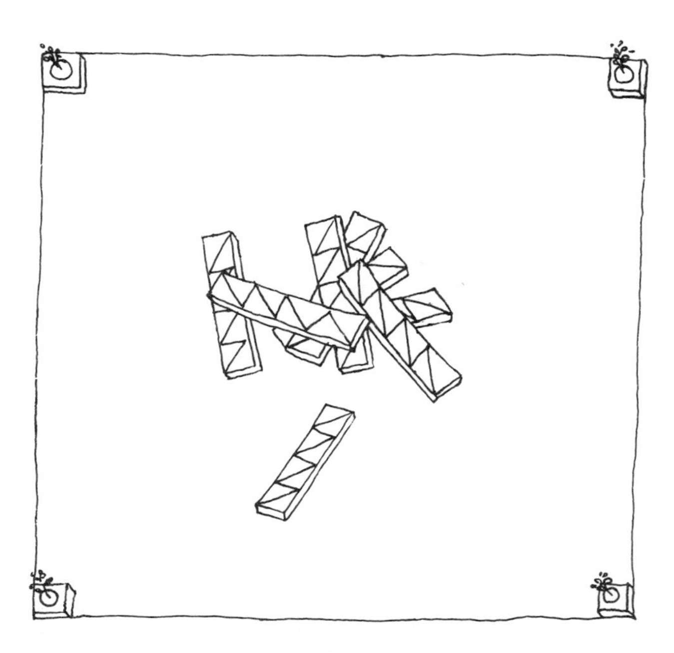

The court is 16 meters by 16 meters. The prince would like to place a 1-meter-by-1-meter fountain in each of the four corners. Can 63 1-meter-by-4-meter tiles fill the rest of the courtroom?

? 2. What is the maximum number of fountains that can be in the corners?

After an hour, Ecco was prepared to give his answer. The prince had watched Ecco's ruminations with great interest.

"Allah is great," said the prince after hearing Ecco's result. "I have one more question. Next to the courtroom is a 10-by-10-meter dining room. Italino Sepe, the best student of Perfecto, has designed 24 1-by-4-meter tiles for that room. I will put four similar fountains in the corners of that room if Allah guides you to a layout."

? 3. Find a design, if possible, for the 10-by-10 room using 24 tiles and one fountain in each corner. If impossible, prove that it is impossible.

31. Oil and Water

The *Newsweek* poll we saw on the flight home found that most people wanted Baskerhound arrested and brought to justice, but they considered him to be more a crackpot than a menace. Not so the government. Frustrated by the inability of the police to arrest him, Congress passed Internal Security Plan I. The legislation permitted the arrest without warrant of people suspected of aiding Baskerhound.

Ecco seemed to pay no attention to this news. He appeared more concerned that celebrity had cut into his windsurfing. Every day, a sack of mail arrived at the MacDougal Street apartment with requests for help from individuals, companies, and governments all over the world. He was glancing through one pile when the return address apparently caught his eye: Baja California.

He made a phone call, and we could overhear the excited Texan drawl of the person on the other end of the line. After he hung up, Ecco turned to Evangeline and me. "A private plane is waiting for us at LaGuardia," he said. "Then a helicopter will take us into Baja. Let's go."

The flights were uneventful, though spectacular. The green-blue Pacific contrasted beautifully with the pale yellow desert sand on the largely uninhabited shore of Baja California. The helicopter dropped us on a hilltop pad a few hundred feet above the sea level overlooking a nearly empty beach. I saw Ecco look wistfully towards the water as the wind whipped up the white caps.

A sunburned man in his late forties, slightly overweight but nonetheless muscular, was there to greet us. "Ross Eliot, originally from Happy, Texas. Mighty pleased to meet you," he said, shaking our hands. "We got us a problem here that only a mathematician can solve. Please come on inside the shed."

He spread out in front of us a map of the area where we had landed. "As you can see," he said pointing at the map, "this part of Baja has a nearly straight coastline. The three red dots on the map are where we will start construction on the offshore oil rigs next month. From south to north, the three sites are called Ace, Big Slick, and Conman. Our problem is to minimize the length of oil pipe we'll need. Pipe is expensive, let me tell you.

"I'll explain the situation a little better. There's a dock built for each oil rig on the shore at the point nearest the rig. Ace's dock is 1000 meters from Ace Rig and 3000 meters south of Big Slick's dock. Now Big Slick's dock is 3000 meters from Big Slick and 2000 meters south of Conman's dock, which is 2000 meters from Conman itself. We want to lay the shortest total length of pipe possible. Our first idea was to put the holding tank at Big Slick's dock and lay pipes directly from the rigs to that holding tank. But we figured out we'd need more than 8000 meters of pipe. Are you following me, folks?"

Ecco and Evangeline nodded. I nodded uncertainly.

Eliot continued, "One of our smart young engineers said we could run pipes from one oil rig to another and then merge the flow from the two rigs. He said that joining the flow from all three rigs produces too powerful a flow. So, to avoid an oil spill, no pipe should carry the flow from more than two rigs. Here's the design we have now. It links Big Slick and Conman with a holding tank at the base of Big Slick. Dang, I know in my gut we can do better. What's your read on the situation, Docs?"

? Design a pipe system that takes the oil to a single holding tank from all the oil rigs without requiring any pipe to carry the flow of more than two rigs. Two pipes may join only at a rig. You may put the holding tank wherever you wish along the shore. The total length of pipes should be less than 7300 meters.

Evangeline clearly liked this problem. "There is a reflection principle at work here, Jacob," she said.

Ecco stared at the ceiling for a moment. Then he said, "Yes, Evangeline. You've got it." A few minutes later, the two had formulated a solution.

"There you are, Mr. Eliot," said Ecco as he presented a pipe design to Eliot in which the holding tank was on the shore but not particularly close to any dock.

"Now, you're thinking like an oilman, Dr. Ecco," said Eliot as he shook Ecco's hand vigorously.

Ecco asked him, "Say, do you happen to know of a windsurfing store nearby?"

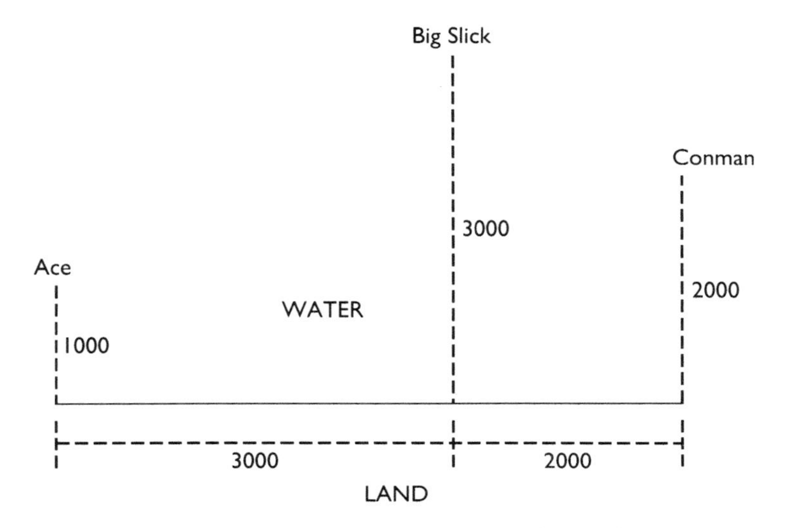

Docks are at the nearest point on shore to each rig. Their position is indicated by the intersection of the dashed lines and the solid line. The holding tank can be placed anywhere along the shore.

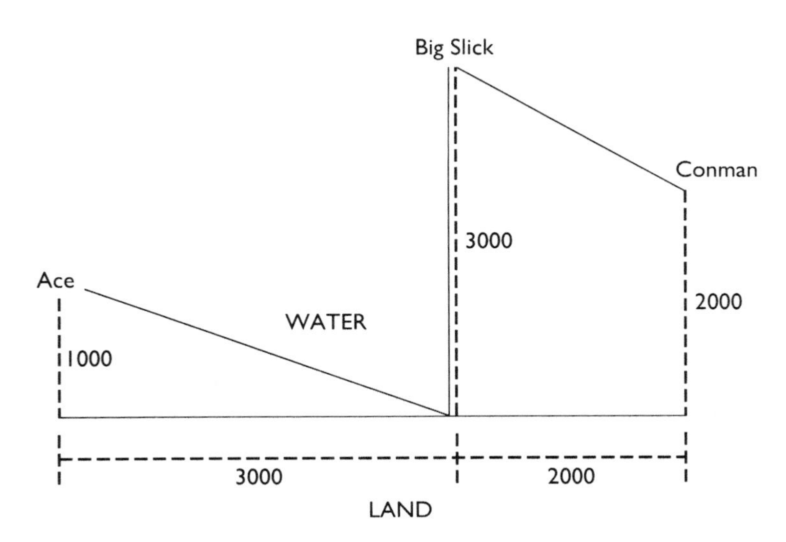

The current design requires over 8000 meters of pipeline. Can you find a design that requires fewer than 7300 meters?

Eliot chuckled. "I thought you two might ask that. You'll find two high-performance Hawaii Thunderbolts at the beach, rigged and waiting for you. Now go bust some waves. You've earned it."

32. Pomp But No Power?

We returned to New York a few days later, refreshed and happy. However, our feeling of well-being did not last long. The day of our return, Evangeline learned that her cousin Michael had disappeared. Evangeline was worried that the police had put him in prison in preparation for deportation. But two days later Michael reappeared in the guise of a delivery boy from a Chinese restaurant.

"I wanted to see what was happening with my own eyes," Michael told us as he laid out the food containers on the table. "Here in New York, the police use their power capriciously, mostly to arrest people whom they consider suspicious, often simply because they have shown up to protest the new laws. Surveillance is still unsystematic. In California, the state police have set up elaborate electronic listening devices, so that any conversation in a car can be recorded."

This bad news was soon followed by more. New legislation suggesting huge changes in the political system was being proposed every day. President Yerrek vetoed much of the more repressive legislation, but votes accumulated to override the veto as legislators outdid themselves in their calls for increased vigilance against nuclear terrorists. As the legislative and executive branches jockeyed for power, the vice president came to consult with Ecco.

Vice President Allan Gear looked younger than his forty-odd years. A decorated veteran, the son of a governor, and the champion of noncontroversial causes, he was careful to exploit opportunities as they presented themselves.

"All laws are up for grabs, Dr. Ecco," he said with a slight Virginian lilt in his voice. "I want to see whether the vice presidency can be made into a real office. I brought my assistant Joseph Brawn to help explain the problem to you. Will you help?"

"I'd like to hear you out first," said Ecco.

"That's a start," said Gear. "The established political consultants avoid the executive office these days. President Yerrek is too liberal, they say."

At this point, Brawn interrupted. "Dr. Ecco, our first goal is to know exactly how much power the vice president now has as a voter in the United States Senate.

"What do you mean by power?" I asked.

"The definition is very simple," Brawn said with a smile. "A person has power if he can change the outcome of some election—from winning to losing or vice versa. As you know, there are 100 senators. The vice president may vote in the Senate only when there is a tie. If we assume that every senator votes either yes or no, then the only tie possible is 50 to 50."

"Fair enough," said Ecco, "so the vice president surely has power. Now, what does it mean for one person to have more power than another person?"

"That's also easy," said Brawn. "A has more power than B if two conditions hold: 1) There is some voting situation in which A can determine the outcome of an election (that is, A can force an outcome of yes or A can force an outcome of no) no matter what B decides to do; and 2) There is no situation in which B can achieve the analogous feat."

Is the vice president as powerful as a senator in a vote?

"So A doesn't have to dominate B in every situation, just in at least one situation?" Ecco asked. "Nice definition. Well, here is my answer."

? 1. What did Ecco answer and why? Assume there are no abstentions.

"Senators sometimes abstain," said Brawn. "What does that do to the power relationship?"

? 2. Would your answer change if abstentions were allowed?

"That's very discouraging," said Gear after hearing Ecco's answer. "Maybe the vice president should have three votes whenever the yeses are within two of the nos. Then I'd have more power than some lousy senator, wouldn't I?"

"With abstentions or without abstentions?" Ecco asked.

"In both cases," said Gear.

? 3. How does Ecco answer?

33. Power Grab

How do you make them stop talking?
Soviet observer at the U.S. House of Representatives in early 1990

A week later, Congress overrode a presidential veto to enact legislation providing for a new national data collection agency. The agency had a mandate "to overcome the threat posed by actual and potential nuclear terrorists by any means necessary." Ecco, Evangeline, Michael, and I gathered in Ecco's apartment, but none of us wanted to talk about the implications of this new agency.

Michael Chang finally began the conversation. "We have entered the stasization stage," he said.

"What does that mean?" I asked.

"Stasi was the nickname of the East German secret police," he explained. "Their most important weapon was the near universal surveillance they achieved. They kept files on one third of all the country's citizens—most of the adult population. When East and West Germany united in 1990, the new government disbanded Stasi. Many former agents became taxi drivers. A common joke of the time was, 'What do you tell a taxi driver when you get into a cab?' Answer: 'Nothing. He already knows where you're going.'"

Everyone except Evangeline laughed. "Stasi was bad," she said in dead seriousness. "But things could be worse here. The government can bring its square miles of sophisticated computers, pilotless spy aircraft, and roving telephone taps to bear in order to achieve a surveillance that Stasi could have never achieved. What an irony! Baskerhound claimed he worked for freedom and now his escapades are leading us to totalitarianism."

"Yes, it's bad," said Ecco, "very bad. But it is not yet time to act. Now, it's time for a trip. Professor, Anatoly Romanevitch has invited you and me to Vosgrad in the Caucasus. Shall we go?"

"Yes, let's," I said, but I was feeling guilty about leaving Evangeline behind.

"Evangeline, will you and Michael be all right?" Ecco asked.

"Yes, we'll keep our heads low," said Evangeline with more hope than conviction.

We took the evening's flight to Moscow and a day and a half later we were in Vosgrad. Located in a mountain valley and hardly changed since the Bolshevik revolution, the town exuded a nineteenth-century charm. Hemp grew wild in the surrounding mountains, explaining perhaps the town's particular appeal to some visitors. Romanevitch arrived a little late to the airport but recognized Ecco immediately.

"Sorry to be late, my friend," he said. "Now that democracy is coming to Vosgrad, the traffic lights have ceased to function, creating traffic jams that even a Los Angeleno cannot imagine. But I have asked you to help me with politics, not traffic."

I saw Ecco wince. The purpose of this trip was to escape politics, not revel in it.

"Democracy is a funny thing," Romanevitch remarked. "You give power to people who have never had it and the first thing they do is to complain that they don't have their fair share—all this, before the first free election has even taken place.

"Here is our problem. We need a fair way to elect our city council. The historic communities are Abinev, Brezmev, Carpow, and Derchev. Abinev and Brezmev have 4000 voters each, Carpow has 2000, and Derchev has 1000. Each community will be allowed one representative. The voters from Abinev and Brezmev want their representatives to have more power than the representative from Carpow. Voters from Carpow would like their representative to have more power than the representative from Derchev. But everyone should have some power. To have power means possessing the ability to change the outcome of some vote. X has more power than Y if there is some voting situation in which X can determine the outcome

The districts of Vosgrad and their populations. Is there a fair way to allocate power among the district representatives?

no matter what Y does, and there is no situation in which Y can determine the outcome no matter what X does."

Romanevitch must have read the same political science texts as Brawn, I thought.

Romanevitch continued, "To satisfy everyone's wishes, we have decided to give different weights to different representatives. What scheme should we use? Oh, by the way, abstentions are not allowed in Vosgrad."

? 1. Would a direct proportional weighting achieve Romanevitch's goals? Give four votes to the representative from Abinev, four to Brezmev, two to Carpow, and one to Derchev.

? 2. Would any other scheme for giving weights to the four representatives satisfy the goals?

Romanevitch took our findings remarkably well. "Dr. Ecco," he said, "suppose I allow you to divide the four districts into subdistricts having any number of voters you determine. Each smaller district will then elect its own representative. However, the council should have only six representatives. Can you devise a scheme that gives each representative from a more populous subdistrict more power than a representative from a less populous one without leaving any representative powerless?

? 3. What do you say?

A Secret Society

34. Odd Voters

The amount of eccentricity in a society
has been proportional to the amount of genius,
material vigor and moral courage
which it contains.
John Locke

We returned to find our country restless and ill at ease. Congress had begun to discuss Internal Security Plan II, which would permit censorship by decree of the National Security Council, universal surveillance at the discretion of the FBI, and the release of government lands for prison sites. Already four military bases had been converted to prisons. Critics of these policies were condemned by self-righteous congressmen, and the FBI started dossiers on them.

"The next step is a show trial," Michael predicted.

Most people kept quiet, including Ecco. If he protested at all, it was with his feet. He accepted a long-proffered invitation from the Omniheuristic Society of Japan. I accompanied him, also glad to leave my country as it continued its short flirtation—as I hoped it was—with totalitarianism. Evangeline, concerned about Michael, stayed behind.

Ecco's Japanese hosts had prepared a full schedule of panel discussions, television appearances, and speeches in secondary schools. Ecco phoned the organization's president, Masayuki Honda, to request to appear at no more than one event—for health reasons, he explained. Mr. Honda agreed immediately: "Your visit honors us greatly, Dr. Ecco-san," he said. "It will be as you wish."

Television crews converged on the Hitachi Research Laboratories outside Tokyo, where a panel was scheduled to discuss the role of omniheurism in the third millennium. Ecco's copanelists, all members of the OSJ, attempted to outdo one another in self-deprecation and praise of Ecco. Ecco, uncomfortable with attention, distrustful of praise, and an unwilling prophet, gave a two-paragraph speech:

> Ladies and gentlemen, good evening. Self-proclaimed experts in science, religion, and policy have predicted the imminent evaporation of the

human mind and human will. Such predictions may be self-fulfilling, if they cajole the world's citizens into an abdication of responsibility. Once human responsibility is lost, so are ethics. Robots are not going to teach us morality.

However, omniheurists know from experience that this prediction is the barest speculation. We know this, because our approach to problems must combine intuition and flashes of insight with the work of computation. The best assurance against this prediction's coming to pass is to nurture the heurist — whether omni or mini — in every man, woman, and child. Thank you.

Our hosts staged a cocktail party in Ecco's honor. At the end of the evening we all walked from the laboratory buildings to the edge of the lake on which they stood. Two swans floated a few hundred feet from the bank, their pale white images reflected in the moonlight.

"My ancestors owned this area and used this lake and surrounding forest as clan hunting grounds," Mr. Honda told us as we admired the stately trees that rose from the opposite shore of the lake. "Nobility still lives here. When Hitachi founded the research laboratories, the emperor donated those two swans."

Ecco accepted a glass of sake, cordially thanked his many well-wishers, then retreated into the shadows of the trees. Soon after I joined him, a man approached us.

"Dr. Ecco, Professor Scarlet," he said with a slight bow, "please call me Mikio Fujisawa. I am the founder of the Oddists."

"Are we supposed to know of you?" Ecco asked politely.

"No, and I'm glad you don't," said Fujisawa with a smile. "Let me explain. We Oddists are a secret society here in Japan. We believe in individual privacy, we support women's rights, and we pursue the study of prime numbers. For the first eight years of our existence, we had no name. But when the famous journalist Hiru Nuraoka heard about us, he wrote a satirical article called 'Strange People.' That is a terrible insult to most Japanese, but we liked the idea, so we decided to call ourselves Oddists. I've come to you this evening, because we need your help. May I pick you up tomorrow morning in your hotel room?"

When we nodded, he said, "Good. We will leave at 7:20 A.M. to take the train to Kyosato."

The next morning we found ourselves traveling due west out of the city. The train rose quickly into the mountains. Fujisawa wasted no time. After we finished the boxed breakfasts of rice, seaweed, and salmon that he had brought for us, he presented his first problem.

"We Oddists make most decisions by voting," he began. "For instance, every night we gather in our meeting hall to vote about whether to go to the Shinto temple. Our voting is completely open. We raise our hands in the style of a New England town meeting. If an odd number of members raise their hands, we go to the temple. Otherwise, we do not. But two weeks ago, our eldest and most respected member, Gaku Ono, had a nightmare filled with spirits, devils, and dragons, all inhabiting the bodies of ferrets. As his dream came to an end, one of the dragons spoke to him clearly and said: 'Privacy, privacy, privacy—even among you—privacy.' After much discussion of the proper interpretation of this dream, we concluded that we must invent a method of voting whereby no member knows any other member's true vote."

"That's simple," I interrupted. "All you need is a secret ballot and a voting booth."

"You do not understand," said Fujisawa, bowing to avoid giving offense. "We still want to vote openly by a show of hands. Our idea was to have our members use their telephones to talk with one another before coming to the meeting hall. The telephone conversations should permit each member to make a certain calculation, so that when the member comes to the meeting hall, that member may or may not vote as he or she originally planned to. However, when the entire group votes, the result—even or odd—should be the same as if each member *had* used his or her original vote. Moreover, each member's original vote must not be evident to any single other member or to any group of other members unless the group consists of all other members."

I saw immediately that no method could ensure that one member would be secure against collusion by all other members. I pointed this out to Ecco: "Following a vote by whatever new method, a group consisting of all members except member B could easily collaborate to deduce the original vote of B. They could compute whether they have an odd or even number of votes without B. If the result is the same as the outcome with B included, then B's original vote must have been nay. Otherwise, it must have been yea."

"Nice observation, Professor," said Ecco. "Let us allow the good Mr. Fujisawa to finish his story."

"There's not much more to it," said Fujisawa. "We have not found a method, but hope that you will."

Ecco nodded slowly and bit a soy cracker that had come with our box breakfasts. "Let me see if I can describe what a solution might look like," he said. "The method is to have three phases. In the first phase, each member B talks to all other members and exchanges some kind of information, but not enough to allow any interlocutor to know B's original intent. In the second phase, member B uses the information from the first phase plus his or her own original vote to compute a vote. In the third phase, all members cast the computed votes."

"Well," I observed, "in the first phase, whatever member B says to any other member B′ must not reveal the original vote of B."

"Yes," said Ecco, "so there should be an element of randomness in what B says. Yet there must finally be order in the randomness. This is the tough part."

? 1. Ecco was eventually able to find a method. Can you? (Hint: if you can't, then read the solution and try to solve the questions below.)

When Fujisawa heard Ecco's solution, he nodded eagerly and said "*Hai*, that is good." But then he seemed to sink deep in thought, looked down at the ground, and said, "Ah . . . Ah . . . " He scratched his head.

"Mr. Fujisawa, is there some problem?" I asked.

If three oddists vote to pray, then they go to the temple. If four oddists vote to pray, then they don't go. Can the outcome of the vote (to go or not to go) remain the same while giving each voter complete privacy?

"It is, of course, a very beautiful solution," he said, "but it may not be practical for our situation. Our numbers have grown large, you see. Your method requires many telephone calls. The telephone company may discover our meeting place. We are a secret society, remember. Is there any adaptation of your method that uses fewer phone conversations, certainly no more than fifteen per member? We really don't ever expect more than five members to collude."

"I can see how it can be done with fewer phone calls," I said.

? 2. How can it be done this efficiently?

Fujisawa was happier with my solution, but thought of a new wrinkle. "Some of our members want to replace two-way telephone calls with one-way fax messages. Using your method would require each member to send many faxes. Since fax machines are slow and fax paper harms the ecology, we'd like to reduce the number. Can you help us still ensure privacy, even if five people collude?"

? 3. Ecco found a method whereby each member needed to send only three faxes. Can you?

35. Polling the Oddists

We left the train station at Kyosato slightly after noon. It was a beautiful clear day and we could see Mount Fuji in the distance. "Seventy thousand people climb Fuji every summer," Fujisawa said as we gazed at the mountain. "That's about 701 a day."

We turned and walked towards the national park. As soon as we entered the gates we saw a huge picnic area with table and restrooms. Families filled the tables and a huge line had formed in front of the ice cream stand.

"It looks very crowded here, doesn't it?" said Fujisawa. "Follow me." We walked a few hundred yards down a trail, over a stream, and up a hill. Finally, half a mile from the picnic area, we sat on a stone bench thoughtfully placed beneath some grand beech trees. Here it was quiet and peaceful. Fujisawa explained his next problem.

"As you can imagine, conducting a prevote poll of the Oddists is more difficult than in other voting situations," Fujisawa began. "It is no good to take a sample, since a single vote can change the entire outcome from even to odd or odd to even. So, we must have a collection of pollsters who ask groups of Oddists what their votes are."

"Excuse me for asking," I said, "but why do you need a poll?"

"We Japanese like to avoid surprises," Fujisawa replied. "Currently there are 60 Oddists (a fascinating number though not odd). Because of our passion for privacy we won't allow a single pollster to interview more than 50 of our members. Oddists will answer pollsters' questions truthfully. If the pollsters also told the truth, then only two would be needed."

"Yes, that's obvious," I said. "For example, the first pollster can interview Oddists 1 through 30 and the second can interview 31 through 60. If both count an odd number of yeas or both count an even number of yeas, then the result is nay. Otherwise, it is yea."

"However," Fujisawa told us, "the pollsters will not be Oddists, because we want to keep our votes private from one another. We think that, at most, one pollster will lie or make a mistake."

"In that case, the minimum number that one would need is four pollsters," I said.

? 1. Why is that the case?

"Right," said Fujisawa. "Nicely done. But now I must ask you the hard question. Are four pollsters sufficient?"

"Easily, if each pollster can report the vote of each Oddist he interviews," I answered.

? Do you see how?

I explained that pollster A could interview Oddists 1–50, pollster B could interview 11–60, pollster C could interview 1–10 and 21–

60, and pollster D could interview 1–20 and 51–60. Each Oddist would then be interviewed by three pollsters. If each pollster reported results on each individual Oddist, then the majority view on that Oddist would necessarily be the correct one.

"Right again, Professor," said Fujisawa. "But we Oddists abhor such an invasion of privacy. At most, we will let each pollster return a number between 0 and 50, not individual votes."

"Perhaps you can use my previous solution and tell each pollster to report the total number of yeas he records," I proposed.

Ecco shook his head. "That won't work."

? 2. Can you demonstrate that Ecco is right?

After showing me the failings of my solution, Ecco began munching on some fried peas.

? 3. There are four pollsters. One of them may lie. Each can interview up to 50 Oddists and report a number between 0 and 50. Whom should each pollster interview? How should the pollster express those results as a number between 0 and 50 to arrive at the right answer?

"The solution uses two tactics," Ecco said. "It exploits the fact that the result we seek is not a total, but merely an even or odd determination; and different responses can be given different weights."

Whom should the pollster poll? What if he lies?

Fujisawa listened carefully to Ecco's explanation, asked several questions, then unfolded a piece of rice paper and filled a page with characters and drawings.

"Have you noticed, gentlemen," Fujisawa asked after he had finished, "how many people we have seen during the two hours we have discussed this problem?"

"I saw two campers on the other side of the stream, but no one has followed us across," I said.

"All national parks in Japan are like this," Fujisawa continued. "The campgrounds are as crowded as subway platforms, but the trails are empty. Few of my countrymen like to be alone."

Contest Puzzle 5: Is it possible to limit each pollster to a number between 0 and 3 inclusive and arrive at the correct answer? If so, show how. If you think not, is it possible to limit each pollster to a number between 0 and 7 inclusive and arrive at the correct answer? If so, show how. If you think not, is it possible to limit each pollster to a number between 0 and 15 inclusive? If so, show how. If not, then just say that it is not possible.

36. The Hokkaido Post Office Problem

We walked back into town. The shops were closed, but we looked at the window displays. I pointed out that many of the summer T-shirts had English writing on them.

"Yes, but look at the English," Ecco responded, pointing to one shirt. It said: "If you look for love in the waste star, you are sure change for a poet."

Fujisawa chuckled. "Yes, the style now is to buy T-shirts that have English words on them. It doesn't matter what they say."

He led us to a waiting van. "I would like you to come to an Oddist meeting," Fujisawa said, "but I'm afraid I must blindfold you." He put padded cloths over Ecco's and my eyes and secured them with Velcro. I felt no discomfort. We drove for about fifteen minutes. When the car stopped, he took off our blindfolds and led us to a mountain meadow surrounded by bamboo trees. It was dark by now and the moon was full.

The Oddists knelt upon large tatami mats in a martial arts rest position, the men on the right and the women on the left.

"Now dressed in an imposing ceremonial kimono, Fujisawa called the meeting to order. All the members fell silent. He spoke in Japanese for about ten minutes; I heard him say Ecco's name several times. Then he paused, and a member stood up as if about to speak. Fujisawa looked in our direction, bowed, and gestured for us to join him.

"Our colleague from Hokkaido would like to ask you a question," he said.

"Dr. Ecco-san," began the man from Hokkaido. "We have only 17 members on the island, but the island is large and we don't all have telephones. Occasionally, we discuss issues by paper mail. Each member sends a copy of his opinion to all other 16 members. On Hokkaido, most letters arrive in one day. However, some do not — some take many days, though they all arrive eventually. The letters may not arrive in order. That is, if A sends a letter to B on day 1 and then another letter to B on day 2, the second letter may arrive first. All letters go through a single central post office. Because of Hokkaido's military importance, we fear that Elder Brother — ah, the government — might read our letters.

"They don't read every letter," he continued. "Instead, a government inspector will arrive periodically, read all the letters in the post office at that moment, and then leave. One of us has found an encoding with the following property: the government inspector must read the opinion of all of our members in order to understand what our discussion is about. Even if he reads one or more copies of the opinions sent by as many as 16 of the members on a particular subject, he will not understand the code, since he will not know the 17th member's opinion."

"So you have to make sure that at all times at least one person has no copies of his or her opinion in the mail?" Ecco asked.

When will the inspector read all the mail?

"Exactly," answered the man from Hokkaido. "For example, here is a protocol for two people. Suppose Shu and Akiko want to communicate. Shu sends to Akiko; when she receives, she sends her message to Shu. At no time would both messages be in the mail at the same time."

"For the problem involving all 17 people, can we assume that some participant, say yourself, starts the protocol?" asked Ecco.

"Yes, by tradition I start all discussions by mail."

"Finally," Ecco asked, "may a participant include any additional message in a letter, say something like, 'I have received messages from these participants'?"

"That is not allowed, since it would violate our code of privacy," answered the Oddist. "In fact, just one other message is possible: the phrase 'all clear.' Now, here are my questions."

? 1. Assume that the government inspector will search the post office only once during the protocol. We want a protocol that will take only three days or fewer, provided all letters arrive within a day. Nevertheless, we don't want our solution to depend on their arriving so quickly. No matter how long a letter takes, the inspector must be prevented from seeing a copy of every opinion. Suppose each participant has 17 envelopes at his disposal in which to send copies of his opinion.

? 2. Suppose the inspector may search twice: find a protocol for eluding the inspector in that case. The protocol should take only five days or fewer, if all letters arrive within a day, but it mustn't allow the inspector to see all the messages in his two looks no matter how long any given letter takes to arrive. Suppose each participant has 17 envelopes in this case, too.

Ecco answered these questions in minutes. The man from Hokkaido bowed deeply after hearing each solution. "Dr. Ecco-san," he said, "recently, we have come to fear that the encoding may not work as well as we had first hoped. Now we think that the inspector may be able to understand the topic of our discussion by reading copies of only 10 messages. So, the question is now the following":

? 3. The inspector may look twice and must read copies of at least 10 opinions to understand the discussion. Each participant has 35 envelopes at his or her disposal. Try to find a solution that takes eight days or fewer, if all letters arrive within a day. Again, the inspector should not be able to understand the discussion, no matter how long any given letter takes to arrive.

After hearing Ecco's solution, the man from Hokkaido bowed deeply once again. "One more question, Dr. Ecco-san, if you please," he said. "A second post office is under construction, and under a special experimental system each citizen will be able to choose the post office through which his or her letters will be routed. The government is hiring no more inspectors. Can we do any better than eight days?"

? 4. The inspector may look twice and needs copies of only 10 opinions to understand the discussion. However, participants can route letters through either of two post offices. The inspector may examine the mail in the same post office on his two inspections, or he may search one post office first and the other post office second. Each participant has 35 envelopes. Try to find a solution that takes four days, if all letters arrive within a day, but which doesn't allow the inspector to understand the discussion, no matter how long any given letter takes to arrive.

37. Joining the Oddists

After Ecco solved these problems for the Hokkaido Oddist, a woman stood up. She spoke in Japanese, but again I heard Ecco's name several times. After exchanging a few words with Fujisawa alone, she addressed the group, again mentioning Ecco. When she finished, many people exclaimed, "Hai!"

Fujisawa explained the situation in English: "Chiharu Narasone nominates Dr. Ecco and Professor Scarlet to become members. To join, you must pass a test. Are you ready for another puzzle?"

"They've been fun so far," said Ecco.

"Very well," said Fujisawa, handing us a sheet of paper. "You have two hours." We sat by a lamp at the edge of the meadow as the Oddists continued their meeting.

"The Oddists revere all things odd," we read from the paper. "To us, the exclusive OR is the beginning of all wisdom. Initiates must learn to appreciate its subtleties."

"Remind me of the definition," I asked Ecco.

"It's very simple," he said. "Think of 1 as representing truth and 0, falsehood. The exclusive OR (xor, for short) of two values is true if exactly one of them is true. So,

0 xor 0 = 0, because neither value is true.

0 xor 1 = 1, because exactly one value is true.

1 xor 0 = 1, because exactly one value is true.

1 xor 1 = 0, because both values are true.

"Let's see what the problem is." We read on: "You are given two sequences 7 bits long; each bit is a 0 or a 1. These are in storage locations X and Y. The only operations you can perform are represented this way:

X←Y,

Y←X

X←X xor Y

Y←X xor Y

"The first two operations copy the contents from one location over the contents of the other. So, X ← Y will write the Y value into the X location leaving Y unchanged. X xor Y takes each bit of X and each bit of Y, performs an xor on them, and outputs the result. Here is an example:

X: 1011010

Y: 0110110

So, X xor Y is 1101100.

Therefore X←X xor Y replaces X by 1101100 and leaves Y unchanged.

"Now, your task is to swap the values of X and Y, that is, after a sequence of operations, X will have Y's old value and Y will have X's old value."

? 1. Can you perform this task using four or fewer of the allowed operations? (If you have trouble, read the solution to see if you can solve the later questions.)

We solved the problem in fifteen minutes, impressing the Oddists. "Gentlemen, we have an advanced test, too," said Fujisawa. "Normally, we offer this test only to members of long standing, but in consideration of the many services you have already rendered, we would like to offer it to you now."

"Please proceed," said Ecco.

"Consider again the setting from the previous problem, except now there are seven storage areas: T, U, V, W, X, Y, and Z," said Fujisawa. "Using a sequence of the same operations as before, you want the following effect:

T should have the original value of Z;

U, the original value of T;

V, of U;

W, of V;

X, of W;

Y, of X;

Z, of Y.

"Dr. Ecco, can you possibly do it?"

? 2. Can you accomplish this with 20 or fewer operations?

I could not even begin to contribute, but Ecco soon turned to me with the solution. "There is something remarkable about this prob-

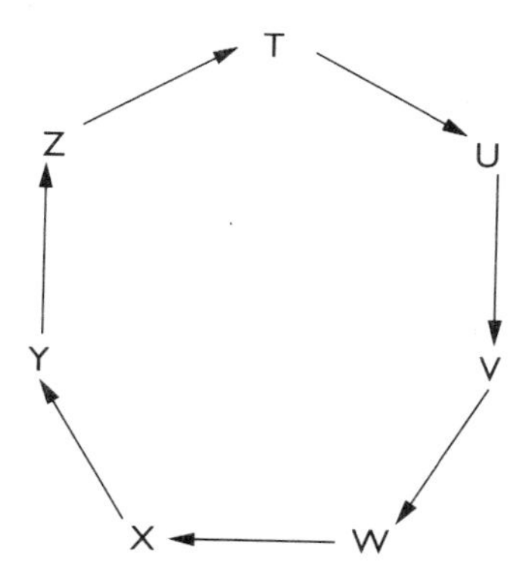

How can Ecco accomplish the seven-way swap using 20 or fewer operations?

lem," he said. "I managed to solve that problem using 18 sequential steps. Yet if we could perform 7 of the allowed operations in parallel, we could finish it all in one step."

? 3. Do you see why?

"Quite surprising," said Ecco. "We can speed things up by a factor of 18 using 7 parallel operations. Doesn't that contradict your intuition, Professor?"

"Yes and no," I said. "I wouldn't be surprised if seven furniture movers could move a large table more than seven times faster than a single mover."

"Are you saying that the parallel approach involves some implicit cooperation?" Ecco asked. "Yes, I suppose you're right. Nice observation."

After Ecco presented the solution to the Oddists, they all rose. "Are you ready for the initiation ceremony?" asked Fujisawa.

We nodded.

Fujisawa and all the Oddists made us a deep bow. We bowed back to them.

"Welcome to the Oddists," they all said in unison.

38. Oddist Summer Training

The Oddist meeting went on, but I fell asleep on the grass. We had been up for twenty-two hours. In the morning, I felt Fujisawa shaking me. "Professor Scarlet, it is time for practice," he said.

"Practice?" I mumbled.

"We Japanese believe in training as a means to perfection," he said. "You have heard perhaps of how our baseball teams prepare for the season."

I shook my head.

"The coach drives them from dawn until night," Fujisawa said. "If he's disappointed with the catcher, for instance, he cuts a square piece of wood and drives twenty-five nails through it. He places this, sharp side up, under the squatting catcher to discourage him from leaning back on his heels. Quite effective."

I could imagine. Groggily, I stood up and washed my face.

Fujisawa explained the Oddists' training exercise. "Each exercise involves a man and a woman—for example, Shu and Akiko," he said. "A judge gives a 17-bit number to Shu and a possibly different 17-bit number to Akiko. Shu and Akiko know in advance that they will be asked one of two possible questions:

1. Does the exclusive OR of the two numbers contain an odd number of bits?
2. What is their sum?

"Shu and Akiko can communicate with one another only by messengered postcard, each postcard consisting of either a 0 or a 1. Each postcard is guaranteed to arrive within an hour of when it is sent. For reasons of privacy, only one postcard can be in transit at a time.

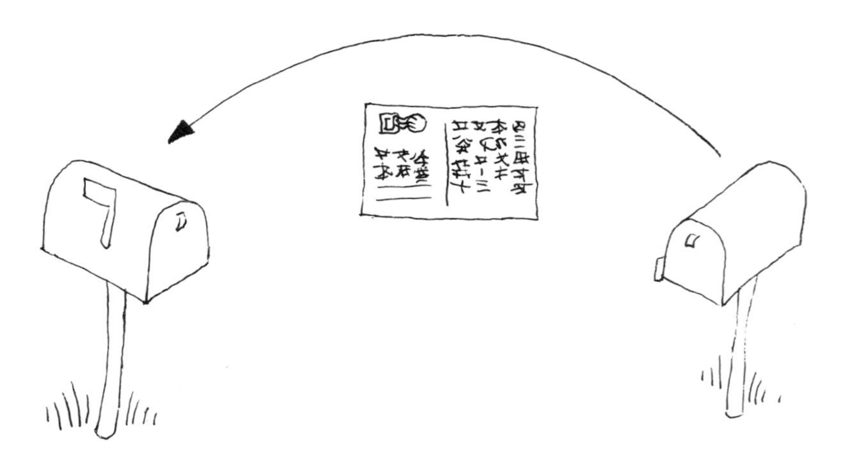

Each postcard takes less than an hour to arrive. How can Akiko send at most 6 postcards with the result that Shu knows the sum of two seventeen-bit numbers within a day?

Whoever figures out the answer sends a postcard to the judge."

"May the two of them meet before receiving the numbers to figure out a strategy?" Ecco asked.

"Yes," answered Fujisawa.

"Well, then, for the question about the odd number of bits resulting from an exclusive OR, all they need is two postcards, including one to the judge." Ecco said.

? 1. Design a protocol that works this fast.

Fujisawa looked down and was silent for a moment. Then he nodded and asked, "Forgetting about the postcard to the judge, can you find a way to compute the sum that requires less than a day and that uses no more than six postcards?"

? 2. Ecco was able to find a solution requiring six postcards satisfying these conditions fairly easily. Can you do as well or even better?

After presenting his answer, Ecco paused, then said, "Fujisawa-san, you make the problems too easy. If you had wanted, I could have found you a 17-hour solution using only five postcards. Not only that, but if there were no need to finish quickly, Shu and Akiko could save a lot of money on messengers. In fact Akiko could send just one postcard if they had 7½ years to play the game."

Again, Fujisawa was silent for a moment. Then he said, "Tell me about your one postcard solution, teacher."

? 3. Can you answer Fujisawa's challenge?

Fujisawa smiled broadly when he heard this solution. "There is no longer any doubt," he said, reaching into his satchel. "This is for you. Please don't open it until you rejoin Dr. Goode in New York." He handed Ecco an envelope on which was written, "For the true Jacob Ecco."

The handwriting was Baskerhound's.

Concentric Conspiracies

39. Plea from a Fugitive

It was only one life.
What is one life in the affairs of a state?
Benito Mussolini

Four days later, back in New York, Ecco opened the letter in front of Evangeline and me in his apartment. It was written in code:

Euue,

Fele ojelek kljefye kujehlk xej ek. Teveq qee eje lze zeje efv I ee lze jenedev ujeeefed. I ee e hejele, cevfehhej, efv eplejleefekl, lzeq keq. Il'k edd ljee, I kehheke.

Bel etene edd, I ee yeedlq ex klehevelq. Se eje qee. Yee kee, Euue, oe kledd zene eeuz ef ueeeef. We eje telz heofk ef e yeee ex klele ueflјed. Ukefy ee ek ef epueke, lze eelzejeleek zene teyef kheflefeeek keejuzek efv keerejek lzefck le lze feo Iflejfed Seuejelq Pdef I. Iflejfed Seuejelq Pdef II ek jeevq le ye efle exxeul. Il oedd eeef leled kejneeddefue efv efenejked uefkejkzeh. Tze xeleje ek udeej xej lzeke oze veje le deec.

If ejvej le xejekledd Iflejfed Seuejelq Pdef II, I oedd kejjefvej. Tzeq oedd tjefy ee le ljeed. Add lze xeulk eje eyeefkl ee. Bel lzeje ek efe keedd efuefyjeelq lzel eeq hejkeeve qee le eul.

U.S. khq kzehk hekkekk e zevvef ljefkeellej oelz e jefye ex 1000 eedek lzel tjeevueklk lze kzeh'k epeul hekeleef. Tze Neleefed Seuejelq Ayefuq zek lze uehetedelq le veueve lzel tjeevuekl. Ix qee eje kcehleued, jexej le dekl Deueetej'k *Dexefke Edeuljefeuk*, heye 81.

Euue, ex qee ve veueve le eul, te hjehejev xej efqlzefy.

BB

Evangeline and Ecco began working on the code immediately with their customary energy. I saw them grow more and more distraught as its meaning became clear.

"Evangeline, I must do something," Ecco said after he had reread the message several times.

"No, Jacob, no," she said, shaking her head. "How do you know it's not just another one of his tricks?"

Turning to me, he said, "You often complain, my dear Scarlet, that I never warn you when I plan to disappear from New York. Well, I will be dropping from sight again for a few weeks. I hope we'll see each other afterwards."

Ecco packed Baskerhound's letter, a book entitled *Inter-Service Rivalries in the U.S. Military,* and some clothes. He embraced us both and left. Evangeline put her face in her hands and sobbed. I put my hand on her shoulder to comfort her.

Two days later, we read that Baskerhound had surrendered to the Baltimore police.

40. The Prosecution Makes Its Case

Everyone, from steelworker to diplomat, had an opinion about Baskerhound on the eve of his trial on charges of nuclear terrorism. By and large the opinions favored severe punishment. Many people agreed with the *New York Post* when, under a photo of Baskerhound arriving at the courtroom escorted by police, there appeared the caption, "Does this man even deserve a trial?" Editorials expressed surprise when Ariana Radan, the renowned corporate litigator, took his case for no fee. "Every man has the right to a competent defense," she explained.

The U.S. Attorney's office selected prosecutor Glenn Green, by every standard the best attorney on the staff. He had won the most convictions overall, the most convictions for difficult cases, had never had a conviction overturned, and routinely sent defendants to prison for maximum sentences. According to the papers, his strategy never varied: thoroughness to the point of tedium. In a low monotone, he would present every fact to support his argument no matter how trifling. While difficult to listen to, he consistently convinced the jury of his sincerity and the justice of his cause. (It was said that jurors who fell asleep during Green's hour-long closing arguments convicted out of fear that they had missed some crucial piece of evidence.)

A few days before the trial, the Director phoned my apartment and asked about Ecco. Evangeline, switching on the speaker-phone, told him that Ecco had gone on a meditative retreat and she did not know when he would return. The Director seemed unconcerned. "Would you and Professor Scarlet care to join us as spectators at the trial?" he asked.

"No, thank you," said Evangeline. "We lead very busy academic lives, as you know."

"Professor Scarlet, are you there?" asked the Director. "Would you mind providing your notes on your captivity with Baskerhound to the prosecutor? They may serve as impeachment material."

"Yes, if you think that is useful," I said hesitantly.

The Director had the notes picked up the next day.

ABC-TV News covered the trial. The day before the trial began, a special broadcast showed footage of the courthouse, the Virginia prison where Baskerhound was housed, and the opposing attorneys in their offices lost in a forest of law books and documents.

"So much drama," I said to Evangeline. "It's beginning to resemble a Stalinist show trial."

"Just as Michael predicted," she said.

In his opening statement, Green told the jury that the prosecution would trace Baskerhound's every step from the time he left Princeton in 1986 to his capture by Federal agents a month before.

As we watched the trial unfold, a parade of witnesses from Uruguay to Thailand took the stand. They described Baskerhound's global wanderings down to the smallest details. Our friend and Baskerhound's electronics technician Kate Edwards recounted episodes of Smartee's three-way interviews: first Smartee and client, next Smartee and Ecco–Baskerhound, then Smartee and client again. This went on for eight full days of testimony. My notes were placed in evidence.

We scanned the papers each day. Society columnist Susanne Post pronounced herself "unsurprised" at the deception by Phillip Andrew Smartee. "Obviously, he was too good to be true," she wrote. Most papers, however, buried the testimony about Baskerhound's omniheuristic exploits in the middle sections, devoting the front pages instead to reports and speculations about Baskerhound's military adventures.

Many jurors dozed off during the interminable testimony of air traffic controllers from around the world as they read from their log books the exact arrival time of Baskerhound's LearJets at one airport or another. Ariana Radan, Baskerhound's attorney, appeared to be drafting contracts for her corporate clients. Had she already given up on the case, I wondered.

Everyone perked up when Green called Captain Whittier, former commander of the nuclear submarine *Groton,* to the stand. He had the unhappy task of describing how Baskerhound had duped him. The following are some excerpts from the testimony as broadcast on television:

GREEN: Captain Whittier, what was the first radio contact you had with the U.S.S. *Freedom,* which Baskerhound then controlled?

WHITTIER: On July 16th of this year, at 0800 hours, they contacted us by radio, claiming they were the submarine tender *Lady Jane.*

GREEN: What did you do then?

WHITTIER: We commenced an authentication protocol.

GREEN: What is that?

WHITTIER: A sequence of message exchanges by which one ship proves its identity to another. Navy regulations require that authentication occur before any normal communication. According to regulations, impersonation is impossible when authentication is conducted properly.

GREEN: But that assertion was false.

WHITTIER: Yes, Dr. Ecco demonstrated later to Admiral Trober that the protocol was faulty. But we have been trained to operate by the book, sir.

GREEN: What happened after you completed the protocol?

WHITTIER: The ship we believed was the *Lady Jane* requested that my ship, the *Groton* surface.

GREEN: And did you?

WHITTIER: Yes, sir, we did, although I remember thinking that the ship had an unusually small profile in the water for a submarine tender.

GREEN: What happened after you surfaced?

WHITTIER: They sent a boarding party of five men in a small boat. They said that they had to conduct an inspection of our safety procedures. They brought on board a device that vaguely resembled a Geiger counter. We showed them around the sub. In the engine room, their device lit up and a red light began to blink as the thing clicked away. The crew became very nervous at that, even though we had checked the engine room for radioactivity levels only the day before. The inspectors asked many questions about operating procedures. They seemed very critical of some of the ways we did things, so we felt compelled to explain ourselves. I know now we gave away our operating secrets.

GREEN: What happened next?

WHITTIER: Because of their apparent displeasure, the inspectors requested an officers' conference. When we gathered in the room, they locked the door, held us at gunpoint, and told us to announce to the crew over the intercom that we were to abandon ship because of a radiation leak. Soon we found ourselves on the U.S.S. *Freedom.* The five men and the rest of their crew boarded the *Groton,* supposedly to decontaminate her.

GREEN: What did you do once on the *Freedom?*

WHITTIER: My fifty men were crowded on the small ship and several suffered sunburn. The Indian Ocean has a merciless sun, sir. The boat had some food and water but no fuel and the ship-to-ship radio had been disabled. Fortunately, we were rescued the same day.

Radan began her cross-examination. "Captain Whittier, you allowed Baskerhound's men to inspect your ship and call all the officers into a room in spite of your suspicions. Wasn't that a violation of Navy protocol, at the very least, if not of a captain's duty?"

Captain Whittier began, "They gave us the intelligence password of the week and—"

Green jumped up. "Your Honor, the People request a conference in chambers. Captain Whittier may be broaching some sensitive issues." The judge granted his request. After a few minutes he, Radan, and the judge returned to the courtroom.

The judge brought the court to order. "For national security reasons this trial will resume behind closed doors beginning tomorrow morning at 9:30," he said.

That was the end of the captain's public testimony. Apparently the very existence of a password used by boarding parties was a secret. After a few days of silence on the issue, the Defense Department issued a statement: "The boarding password of the week was broadcast in the clear over the standard naval radio frequency in the Diego Garcia area the day before the *Groton* was captured. Source is unknown."

The trial reopened to the public two days later. The prosecution called Tim Crews, the pilot who had spotted the spy ship *Freedom* and rescued the drifting crew.

GREEN: Lieutenant Crews, how did you locate the ship?

CREWS: I just saw it.

GREEN: And what did you do when you saw the ship?

CREWS: I radioed the nearest cruiser group.

GREEN: You saved many lives, Lieutenant. No further questions.

Crews nodded smugly. Ariana Radan's cross-examination was brief.

RADAN: Lieutenant, were you flying on a mission when you saw the ship?

CREWS: Yes.

RADAN: What was your altitude?

CREWS: Standard spy-cruising altitude. Seventy thousand feet and above.

RADAN: From that altitude, you recognized the *Freedom*? Perhaps you read its name on one of its life preservers?

Some members of the jury chuckled. Crews shifted nervously in the witness chair. "Well, not exactly," he said. "My receiver picked up the signal of the homing transmitter in the ship. I descended to investigate and found the vessel drifting—"

Again, prosecutor Green jumped out of his chair and moved that the trial continue behind closed doors, but it was too late. Within a few hours, a spokesman for the Defense Department announced at a press conference: "DOD installs transmitters on certain naval platforms that broadcast location coordinates continuously. The U.S.S. *Freedom* possesses such a transmitter. In has been in good working order for the life of the vessel."

For his last witness, the prosecutor called Nigel Williams.

GREEN: Mr. Williams, were you with Dr. Baskerhound during the time he controlled the submarine *Groton*?

WILLIAMS: I was indeed.

GREEN: Were you present when he made his nuclear threat to the world?

WILLIAMS: Yes, sir.

GREEN: Can you describe his emotional state at that time?

WILLIAMS: He was a bloke a little on the excitable side, if you get my meaning. I sometimes feared he might blow the whole planet to smithereens even if he got what he wanted.

GREEN: So, you think he actually intended to launch nuclear weapons against major world cities?

WILLIAMS: Yes, if he hadn't been stopped in time.

Ariana Radan moved forward in her seat as if about to raise an objection, but said nothing. Green told the witness, "Thank you, no further questions."

Ariana Radan limited her cross-examination to four questions.

RADAN: Mr. Williams, you have given this court under oath your real name, have you not?

WILLIAMS: Of course I have, ma'am.

RADAN: But at least once in the past, you went by the name Hanson and posed as a treasure-hunter from Virginia. Still another time, you wore the hat of a Sotheby regular. Isn't that true?

WILLIAMS: Yes.

RADAN: You were trying to deceive people when you took those aliases, weren't you?

WILLIAMS: You could say that.

RADAN: Were you lying then or are you lying now?

WILLIAMS: I'm telling the truth now. I swear it.

RADAN: Additional swearing is not necessary, Mr. Williams. No further questions.

Radan glanced significantly at the jury as she returned to her seat.

The press exploited all angles of the story. They printed the testimony of Nigel Williams verbatim in family magazines. They went on to paint an apocalyptic vision of what could have happened had Ecco and the Navy Seals not disarmed the *Groton.* There was little sympathy for Captain Whittier. *Newsday* ran a cartoon showing a child pointing to a toy boat and saying, "Dat's an aircraft carrier." A caricature of Whittier nodded his head and said, "Yes, sir." Cloe Anne Bennet's article for the *New York Times Sunday Magazine,* "Baskerhound's Voyage: A Slip Through the Maze of Government," began this way:

> When the prosecution rested in Dr. Benjamin Baskerhound's trial on Thursday, most observers expressed shock at the many blunders of the U.S. military. The gaffes allowed Baskerhound to seize the submarine *Groton.* The testimony revealed how a sham omniheurist gained entrée to the highest levels of government and how he persuaded them to adopt a faulty naval security protocol. There followed uncoded broadcasts of a crucial security password. Finally, a gullible officer surrendered his ship

to a tiny vessel that could not possibly have been the submarine tender it claimed to be. The result was to allow Baskerhound to take the submarine and practice unprecedented nuclear blackmail. Nigel Williams testified that Baskerhound was closer to pushing the button than anyone knew.

"The case against Baskerhound seems open and shut," I said after finishing the article.

"Funny you should say that," said Evangeline. "I now view him as a marionette rather than as a terrorist."

? Please stop a moment. How would you defend Baskerhound if you were Ariana Radan?

41. Friends in High Places

Evangeline and I watched on television as Ariana Radan began the case for the defense on a Monday morning. Her opening statement was enigmatic to say the least: "Ladies and gentlemen of the jury, to understand this case, you must look beyond Benjamin Baskerhound. Just how far beyond will be evident shortly." With that, she stepped away from the jury box and faced the judge. "As my first witness," she said, "I call Dr. Jacob Ecco." There was a murmur in the courtroom. The judge banged his gavel repeatedly and the television station quickly cut to commercials. I glanced at Evangeline.

"I'm as surprised as you are, Professor," she said.

When coverage resumed, Radan posed her first question.

RADAN: Dr. Ecco, what is your profession?

ECCO: Omniheurist.

RADAN: And what does that mean?

ECCO: Literally, it means a solver of all problems. In my case, those problems must submit to logical analysis.

RADAN: Do you have formal training in this field?

ECCO: I have a doctorate in mathematics from Harvard, but that was only partial preparation.

RADAN: What about practical training?

ECCO: Yes. In well over a hundred cases, some documented by my colleague Professor Scarlet, I have applied omniheurism to problems ranging from finding spies to designing buildings to locating submarines.

RADAN: May it please the court, the defense offers Dr. Ecco as an expert in his field, the field of omniheurism.

Glenn Green nodded and said, "No objection, although I find such a form of expertise to be peculiar. However, in deference to the many achievements of Dr. Ecco, I will accept it, Your Honor." The examination continued.

RADAN: Dr. Ecco, you've heard and understood the evidence presented by the prosecution, have you not?

ECCO: Yes, I have.

RADAN: Assuming it is all true, what conclusion do you reach?

ECCO: It is unlikely that Baskerhound acted on his own.

RADAN: Can you elaborate?

ECCO: The simplest theory that fits the facts is that Baskerhound had help from one or several people in government who hold power over the Navy, the Air Force, and the intelligence services.

Green sprung to his feet: "Objection! Your Honor, the witness is speculating. I move to strike the answer from the record."

The judge agreed. "Motion granted. The jury is instructed to disregard the witness's last answer. Dr. Ecco, you are here to give us the benefit of your expertise, not jump to speculative conclusions."

Ecco did not respond or even look at the judge but continued to stare straight ahead.

RADAN: Dr. Ecco, please explain to the court the reasons for your conclusions.

ECCO: First, circumstantial evidence. Baskerhound was heading towards Moher before the U.S.S. *Freedom* even was lost. That could have been good luck, but it would have been extraordinarily good luck. Second, recall that someone broadcast the boarding password the day before Baskerhound seized the submarine. Either someone committed an unprecedented blunder or someone made an intentional contribution to Baskerhound's plan.

GREEN: Your Honor, once again the witness is speculating. I move that this answer too be stricken from the record.

JUDGE: Motion granted. Dr. Ecco, please refrain from giving a theoretical disquisition.

RADAN: Is there any other evidence you would like to bring to our attention?

ECCO: Once Baskerhound had taken the *Freedom,* the National Security Agency had only to send a plane within a thousand miles of the ship to receive and decode the signal from the *Freedom's* hidden transmitter. That is how Tim Crews found the ship when it was drifting. In fact, the NSA could have recovered the ship within days of its disappearance. Everyone knew that the ship had disappeared off the Cliffs of Moher. The NSA could have sent planes to scan a radius of a few thousand miles centered at the cliffs. There can be no excuse for such inaction, nor was there any excuse for failing to capture Baskerhound in the eight months between the capture of the *Freedom* and the capture of the *Groton.*

The court was in an uproar. Green looked as bewildered as the jury.

RADAN: Where does this lead you, Dr. Ecco?

ECCO: Here I must speculate: Baskerhound had help from people with access to NSA equipment and also from an unknown source that knew a crucial naval password. So, someone in the Navy was also involved. Historically, the NSA and the Navy are rivals. No person within either organization could have prompted such bizarre behavior from the other, at least not through official channels. I assert that Baskerhound's accomplices must have had power over both organizations.

GREEN: Objection, Your Honor. This is sheer speculation.

JUDGE: Overruled. The witness has described it as such. You may cross-examine later, if you wish.

RADAN: You may cross-examine now, Mr. Green. Your witness.

GREEN: Dr. Ecco, have you ever mentioned this conspiracy theory of yours to anyone?

ECCO: No, I haven't.

GREEN: In Professor Scarlet's notes during his imprisonment by Baskerhound, you are portrayed as a willing helper. Do you recall this passage from those notes? "I was relieved to see that he was in good health. Almost too good, I thought. Unlikely as it seemed, I wondered if Ecco was a prisoner or an active participant in 'Club Baskerhound.'"

ECCO: Yes, I do.

GREEN: Do you realize that your testimony may be seen as a mitigating circumstance in Baskerhound's crime?

ECCO: Yes, I do.

GREEN: Then why shouldn't the jury believe your testimony is motivated by your friendship with Baskerhound, an alleged nuclear terrorist?

Murmurs could be heard through the courtroom. The judge restored order.

RADAN: Objection, Your Honor. The prosecution is badgering my witness.

JUDGE: Sustained. Dr. Ecco, you need not respond.

ECCO: Yes, thank you, Your Honor. I will respond. The evidence speaks for itself, if we just listen. The transmitter should have made it easy to find the ***Freedom***, yet the ship was never recaptured. Only once in the history of the U.S. Navy has a password ever been broadcast in the clear. That was the day Baskerhound captured the *Groton.* I was inevitably drawn to this conclusion: Baskerhound had help—in high places.

Green looked pale, but he tried one last tactic. "What is your motivation, Dr. Ecco?" he asked. "What you are saying borders on treason."

Ecco paused and thought for a moment. Then he replied softly, "To regain our liberties."

The courtroom exploded. The judge hammered repeatedly with his gavel, but he had to threaten to clear the courtroom before everyone settled down. Finally, he spoke: "This court will recess until tomorrow."

We expected Ecco's testimony to dominate the evening news, but, remarkably, there was no mention of the case. Later in the evening, President Yerrek made the following announcement: "My fellow citizens, it is alleged that Baskerhound had help from officers of the federal government. This is a serious charge and I would disregard it, were it not for its source. I say this to all citizens: if you can bring light to this matter, step forward now. You owe it to your country and possibly to your liberty."

But that was it. The reporters did not follow up. Silence reigned.

? What is your hypothesis?

42. Nightly News

Ecco did not return home that night. Now that his testimony was over he had no reason to hide, so we worried about his absence.

The phone rang. Evangeline turned on the speaker-phone.

"This is Cloe Anne Bennet of the *Times.* May I please speak to Dr. Goode or Professor Scarlet?"

"We don't give interviews," Evangeline said abruptly.

"I don't wish to interview you," said Bennet, "but I would like you to come to my office."

"But it's 11:30 P.M.!" I exclaimed.

"It is imperative," she replied. "I'm on the ninth floor of the *Times* building. Be sure to come up the elevator alone."

Something about the urgency of her tone made us willing to brave the rain, but before we left, Evangeline made a brief phone call in Chinese. Then she slung a camera bag over her shoulder and we stepped out into the night.

When we entered the building, we phoned up to Bennet and took the elevator. There were no other passengers. Evangeline pushed 9, but at the sixth floor the doors opened. A woman stood before us.

"I'm Cloe Bennet. I must see you," she said. "Come with me."

Evangeline and I glanced at each other. We followed Bennet into a small wood-paneled study. A tape recorder lay on the table.

"You are probably wondering why I dragged you out of your apartment in the middle of the night, and why I intercepted your elevator on the way up," she said.

"We are indeed," said Evangeline.

"I am a *Times* reporter, that is, a professional observer," Bennet continued. "But Dr. Ecco's testimony has made me a participant in a way that I never dreamed possible. I need your help to free me and possibly to save the country. Ecco's testimony suggests that Baskerhound may have been entrapped. But Baskerhound's relative innocence is not the main issue. His guilt is clear no matter what moves Radan pulls. No, the main question is who the accomplices might be, if any. If they are as high up as Ecco suggests, their actions constitute a conspiracy of unprecedented danger in the history of the United States. Yet you didn't hear anything about it in the nightly news, and you won't read about it in the *Times* tomorrow. My editor has taken me off the story."

Her eyes moved slowly from Evangeline to me and back again, as if to see whether we appreciated the implications of what she had just said.

She continued. "This afternoon, not an hour after Ecco's testimony, my editor called me into his office. 'Bennet, I am reassigning you to City Hall,' he said. 'Don't give me any lip or you are fired. Good day.' Now, this editor is tough, but he's no bluffer. I was speechless with rage. I walked out of the office back to my desk, resolved to take the story to another newspaper. My desk phone rang. A man whose voice I did not recognize told me the exact whereabouts of my son, the names of his friends, and his favorite ice-cream flavor. 'Don't do anything brave,' he said. I was shaking when I hung up the phone.

"About an hour after the phone call came, a messenger brought me a package. It contained a single cassette tape and a note which I could not understand. The tape is truly frightening."

We looked at the note. Evangeline began working on the code.

> Tzek lehe oek jeuejvev ef e keedd keleeleef jeee ekev tq eeetejk ex lze Neleefed Seuejelq Ceefued. Bq ef edv hjekevefleed vejeulene, lze jeee ek edoeqk teyyev. Heje eje keee epuejhlk ex uefnejkeleefk lzel zene lecef hdeue lzeje. Tjq le uefneq lzek uekkelle le Peed Sennq el ABC Neok. He oedd cfeo ozel le ve oelz el. Tjetej

After decoding the message, Evangeline reread it several times. Then she closed her eyes and took a deep breath. "Give us a blank cassette and the package in which the tape was delivered," she said. Bennet complied and Evangeline put the blank cassette in the package. Evangeline scribbled a note in Chinese and put it with the real cassette into her camera bag. She stuffed the encoded note into her pocket.

We returned to the street, more than a little jumpy. As we passed an alley, I saw a figure stagger towards us. He avoided me but bumped into Evangeline. She said nothing. It took me only a few seconds to realize that she no longer held her camera bag. I looked behind us, but the drunk had disappeared.

Evangeline put her hand on my arm and said, "I'm all right, Professor. Let's just keep walking."

We hailed a taxi on Fifth Avenue and I told the driver to take us to my apartment. He calmly turned his head and pointed a gun at us. It was Williams.

"What do you want?" I asked, trying to remain calm.

"Only the package," he said.

Evangeline handed it to him.

"Now get out," he said, waving his gun at us.

We didn't need to be told twice. He sped away, leaving us standing at the curb. The cab had no license plate. I hailed a second taxi, which took us to Washington Square without event.

In my apartment, I turned to Evangeline. "What happened to—"

Her look stopped me short. "Let's go to sleep," she said. "It's been a long day."

To my great surprise, I did fall asleep. When I woke up, I turned on the news. The top story was that Benjamin Baskerhound had escaped.

43. Television Treason

We stayed tuned to ABC, but there was nothing about the trial all day, not even any details about how Baskerhound might have got away. Around 4:00 P.M., my fax machine printed out five lines on a single sheet of paper.:

Reversals.
Bockhf jaaj.
Dignheg C xz bf mz nerlyzm iwytdq.
Clbvqb kk mxuv.
Aaltpz wp jwosxsyni.

Evangeline was still working on the code when the evening news began. The first item was a live speech by the Director in which he said, "I've been instructed by President Yerrek to implement Internal Security Plan II for the duration of this crisis. Baskerhound is known to be armed and is extremely dangerous. To avoid leaking information that may lead him to capture another ship, all newspaper articles and TV news broadcasts will have to be approved by the security agencies before they are disseminated. This is a necessary measure. Resistance will be considered treason. Thank you for your cooperation."

Suddenly, the image of the Director's face froze on the screen. We heard two voices. The second was clearly the Director's.

VOICE 1: Sir, Smartee is a front.

DIRECTOR: I know.

VOICE 1: What should we do?

DIRECTOR: Get him clients.
(Pause)

VOICE 1: Sir, Baskerhound has seized the *Freedom.*

DIRECTOR: Good. Soon he'll pose the threat we need.
(Pause)

VOICE 1: Sir, Rumtopo reports that Baskerhound is back in Uruguay. Goode and Scarlet are his prisoners.

DIRECTOR: Biding his time, that Baskerhound, isn't he? Maybe we can get him to move. Arrange for a courier to Williams.
(Pause)

VOICE 1: Sir, the *Freedom* is approaching Diego Garcia.

DIRECTOR: Get me Smoot . . . Chuck, hi. The hound has caught the scent. Give him the word. Bye.
(Pause)

VOICE 1: Sir, the submarine crew has been forced onto the *Freedom.*

DIRECTOR: Perfect. It's time to find the ship. Who is close enough to make the hit?

VOICE 1: Only Lieutenant Crews, Sir.

DIRECTOR: Oh, Mr. Hot Dog, himself. Get him on the phone . . . (Pause) Crews, you know who this is don't you? . . . Good. It's time to find the *Freedom.* Take off on a standard mission. You should just happen to find the ship . . . (Pause) Yes, let the hidden transmitter guide you. Understand? . . . Good. Oh, and Crews . . . don't try to grandstand.
(Pause)

VOICE 1: Sir, Baskerhound has just threatened to nuke the world in five days unless he is given $40 million in supplies for the next year, and we clear out of Antarctica.

DIRECTOR: Good. Get me Quinn of the FBI . . . (Pause) Chet,

Internal Security Plan I . . . send it to Senator Phelms. He will get it done.
(Pause)

VOICE 1: Sir, Captain Whittier of the *Groton* sent an urgent message to Trober. He thinks Baskerhound has all codes necessary to launch the missiles.

DIRECTOR: Nonsense — impossible.
(Pause)

VOICE 1: Sir, Baskerhound has just broadcast the launch codes in the clear.

DIRECTOR: Bastard. Now we'll have to stop him.
(Pause)

VOICE 1: Sir, Baskerhound has not been found in Smartee's house.

DIRECTOR: Excellent. It's the perfect excuse. Ask Phelms to bring Internal Security Plan I to a vote.

There was a pause. A clear third voice began speaking: "Ladies and gentlemen, draw your own conclusions. This is Paul Savvy, ABC News, Washington."

The screen went blank, for a moment, then showed an eagle on the background of a U.S. flag.

44. Peirce's Beanbag

Evangeline, her cousin Michael, Ecco, and I met for tea in Ecco's apartment, nine days after his testimony. While Evangeline and Michael practiced a violin duet and Ecco read the *American Mathematical Monthly,* I sat down and read Cloe Anne Bennet's detailed account of the events. Her story was headlined "Rise and Fall, Fall and Rise."

A short month ago, hardly a person had a kind word for Benjamin Baskerhound. At that time, most of the nation's citizens admired and respected the Director, even if they found his manner imperious. Today, many view the Director as the leading conspirator in the most dangerous plot in the history of the Republic.

Officially, the Director—his name is still a secret—directed a tactical staff group at the Pentagon. Unofficially, he wielded power in many ways comparable to that of the Joint Chiefs of Staff or even the president.

According to recent disclosures by investigators, he used the normal agency command chains only occasionally. Most often, he issued his orders directly to people who knew him only by his voice or title. He demanded complete obedience and absolute secrecy. Officers who hesitated, even for a moment, found their careers destroyed. Those who dared violate the code of secrecy saw physical harm come to themselves or their families. According to his former confidant, Mr. Kevin Mealy, the Director was behind the downing of National Security Agency pilot Frank Hart's plane three years ago, after Hart refused to aid drug traffickers by reporting the movements of Drug Enforcement Agency planes off southern Florida.

The Director issued the command to send the spy ship *Freedom* to Moher and then informed Baskerhound by anonymous letter of the ship's position. The letter also provided Baskerhound with an escape route. Then the Director ordered away all ships or planes that might detect the hidden transmitter of the captured spy ship. Secretly, however, four planes continued to track the ship, reporting every move directly to the Director and his private army.

The Director knew that the spy ship posed no significant threat. He knew from his agent Nigel Williams (whom he had planted as one of Baskerhound's chief assistants) that Baskerhound had engaged ex-submariners as part of his staff.

On June 11, the Director heard that Admiral Trober's staff had asked Smartee to design an authentication protocol. Knowing that Baskerhound would exploit this opportunity for his own ends, the Director had his research team examine the protocol to see if Baskerhound had left a trapdoor. The team confirmed his suspicions.

A month later, the U.S.S. *Freedom* approached Diego Garcia, site of the large U.S. naval base in the Indian Ocean. Through his agents at the Department of the Navy, the Director learned that no submarine would permit entry unless the boarder gave a special password. The Director ordered spy plane pilot Charles Smoot to broadcast that password in the clear. Smoot's plane dropped a parachute containing a transmitter with the recorded password. The transmitter sent its single message just before disappearing into the ocean.

On July 18, two days after Baskerhound captured the submarine *Groton*, Baskerhound made the first of several threats over the ship's radio. Nuclear panic quickly gripped the nation. The threats gave the Director a pretext for pressuring Congress to grant sweeping powers to security forces, including the right to impose preventive detention of suspects. The detention prisons filled up quickly. The Director continued to feed the hysteria, hoping that he could eventually introduce harsher laws. Then came the first big surprise.

Baskerhound had cracked the launch codes on the missiles. The Director and the Navy enlisted Ecco's help. Ecco, together with the Navy Seals, located and recaptured the submarine on August 15. Mysteriously, though, Baskerhound escaped.

According to Mealy, the Director had arranged Baskerhound's escape from Smartee's mansion and had had Baskerhound followed. In the meantime, he pressured Congress to pass even more restrictive laws, including strong censorship provisions, dubbed Internal Security Plan II "should the need arise." Baskerhound remained at large until September 22, when he gave himself up. "We are all prisoners now," he said when he surrendered.

The trial began two weeks later on October 2. It was the Director's idea to make the proceedings public, according to Mealy. He wanted a clean and public conviction to justify the security measures he had forced upon Congress.

Ecco's testimony was the second big surprise. Without contesting Baskerhound's guilt, Ecco suggested that Baskerhound had been helped. His speculative theory was confirmed within thirty-six hours, thanks to the bravery of ABC newsman Paul Savvy and especially of a previously unknown police officer named LeRoy Pete.

As pieced together from hundreds of witnesses by this reporter, the events following Savvy's broadcast occurred in this order. After his speech, the Director went directly to a party at the home of Senator Phelms. The Director was in high spirits. When he heard what Paul Savvy had done, he boasted to his fellow party-goers, "Internal Security Plan II will give me the power to pulverize those news showboys. Within a year, I will have imposed a truth so total that no sane person will consider questioning it. The few doubters will be labeled deviants."

Witnesses say that some of Senator Phelms's guests applauded these remarks, but the nation at large was not in tune with such sentiments. Every person in the country old enough to understand the news was discussing what had happened. Some viewed Savvy as a traitor, others thought the tape was a fake. But most thought the tape recording confirmed Ecco's allegations.

Officer Pete approached the Director as he left Senator Phelms's party that evening. "Mr. Director," said Pete, "did you set up Baskerhound?"

The Director turned to Pete, then gave a signal to his bodyguards. Two of them shot Mr. Pete at point-blank range, killing him instantly. This was too much for the other officers who had been stationed around the Phelms residence during the party. They rushed at the Director's entourage, their pistols drawn. In the ensuing gun battle, the Director was shot in the thigh. His bodyguards surrendered to the police.

Hoping for a presidential pardon, Kevin Mealy was the first to confess the wider plot and the aid to Baskerhound. He said that the Director, through his agent Donaldo Rumtopo, proprietor of a casino in the Uruguayan resort of Punta del Este, had given Baskerhound hints, such as when to go to Moher. Baskerhound had trusted these tips, even though he never learned their source. Finally, Baskerhound's escape was a lie. Baskerhound was still in the custody of the Director's men who had given him dangerous antimemory drugs.

After Mealy's remarks were made public, others stepped forward with confessions. All told, the Director had forced literally hundreds of government and military personnel to lie, steal, and even murder to further his own ambitions.

Within seventy-two hours of Ecco's televised testimony, the Director and most of his top-level assistants were behind bars. Now voluble witnesses revealed a huge organization, a secret government within the intelligence agencies. According to many observers, our country had come dangerously close to becoming a totalitarian state.

Because of the harmful drugs he was forced to ingest, Baskerhound has become an incoherent psychotic. Mr. Green granted Ms. Radan's request that Baskerhound be treated as insane rather than as criminal. On October 18, he was committed to Bellevue, where doctors say they will attempt to implant a bioelectronic device into his brain. This will be the first time such a device has ever been implanted into a human being. Chief surgeon Sally Anya refuses to speculate on the results. She made only one comment: "Our implantation device works by 'waking up' neurons near the device. In theory, those neurons will 'wake up' more neurons in an ever-spreading ripple. The cure takes several years. At the end of that time, Dr. Baskerhound could in principle have a more active brain than ever before in his life. We can only hope."

Bennet omitted her own important role in giving us the tape and the note. Michael had delivered the tape to Savvy. He had posed as the drunk who bumped into Evangeline the night we received the tape from Bennet.

"Unusual sort of adventure for you Ecco," I said when I finished the article.

"Yes. Green was right to object to my testimony," Ecco replied. "It wasn't a logical analysis, simply an educated guess. I couldn't point to a naked emperor and proclaim his nudity. Instead, I had to show that the emperor had exposed his thigh. From that point on, I used abduction."

"You mean, you kidnapped somebody?" I asked.

"Not exactly," Ecco answered with a smile. "I used Peirce's notion of abduction. You're not familiar with Peirce's beanbag, are you?"

"No, I'm not," I admitted.

"In 1878, Charles Sanders Peirce wrote the following to explain the different forms of inference:

Deduction:

Rule: All the beans from this bag are white.

Case: These beans are from this bag.

Result: Therefore, these beans are white.

Induction:

Case: These beans are from this bag.

Result: These beans are white.

Rule: Guess that all the beans from this bag are white.

Abduction:

Rule: All the beans from this bag are white.

Result: These beans are white.

Case: Guess that these beans are from this bag.

"Whereas my normal method of inference is deduction, in this case I chose abduction, reasoning as follows:

> *Rule:* With help from high U.S. military officials, Baskerhound could seize a spy ship, control it, and use it to steal a submarine.
>
> *Result:* Baskerhound did in fact do these things.
>
> *Case:* Guess that high U.S. military officials helped Baskerhound.

"It was only one possibility among many. Foreign powers might have helped him, but they would have had too much to lose. I also guessed that demonstrating the plausibility of a conspiracy might do two things. First, it might uncover evidence such as the cassette sent to Cloe Bennet. Second, people would understand that evidence in the framework of a conspiracy. With that, we were lucky. Thanks to LeRoy Pete's bravery, the Director was stopped. It is tragic that lives were lost, but Internal Security Plan I had already claimed many more."

"If it hadn't been for Pete, do you think people would have resisted?" I asked.

"I wasn't sure at first," said Ecco, "but when I saw the angry outpouring against Congress for allowing the Director to influence policy so forcefully, I felt better. Americans instinctively mistrust government." Ecco took a bite of an oatmeal cookie, then continued. "I'm astounded at Baskerhound's insight into all of this. The hidden transmitter was clearly the critical piece of evidence. It should have led to the immediate recovery of the ship. Mere incompetence could not explain the delay."

He obviously had more to say, but the doorbell rang. The caller was Dr. Sally Anya herself. A woman in her thirties, she had an upright bearing and a straightforward manner. After introductions, she addressed herself directly to Ecco. "Dr. Ecco, we did not divulge to the press all the complexities of this case. We knew that the notion of implantation was familiar to the public, so that was what I told Bennet. However, things are not quite so simple."

"Please explain," said Ecco, offering her a chair.

Dr. Anya sat on the edge of the seat and leaned forward towards Ecco as she spoke. "New advances in neuroanatomy have identified some twenty main electrical conduits, called neurocables, connecting nine major regions in the brain. At this moment, all nine regions in Dr. Baskerhound's brain are asleep. It has long been known that a region can be awakened by direct stimulation. But using direct stimulation on more than three or possibly four regions is medically contraindicated. In plain English, it is probably fatal.

"A recent paper by John Eliot suggests that the neurocables can be used to achieve indirect stimulation of brain regions. Using data from rat experiments, he hypothesizes that if a region X is surrounded by at least three awake regions for a year, then region X will wake up. It is well known that a region will stay awake once it wakes up. Our goal, of course, is to reawaken all nine regions of the brain in as few years as possible. My plan is to perform direct stimulation on three regions next week. Once the reawakening process begins, further direct stimulation is contraindicated."

I guess that means he would die, I thought.

"Let me see if I understand," said Ecco. "If you started by stimulating B, D, and I, then after one year B, D, I, E, C, and H would be awake. After two years, A and F would also be awake."

"Yes, exactly right," said Dr. Anya. "Now, the question is: Which regions should we start with so that we can finish in under five years? Are you proposing B, D, and I?"

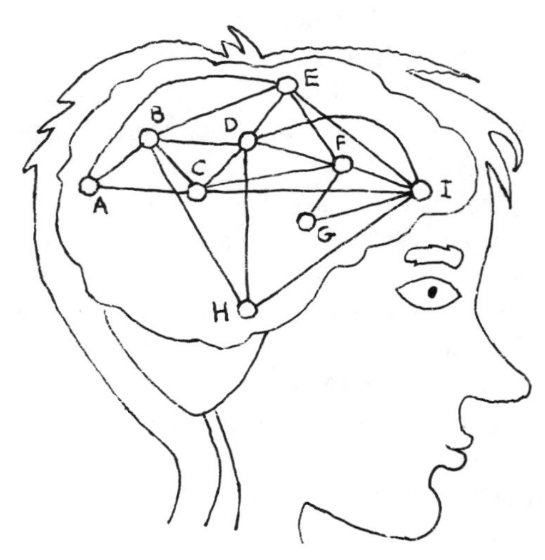

Brain regions A–F and connecting neurocables.

? 1. Ecco was able to find a solution that reawakened Baskerhound's brain after four years starting with direct stimulation of three regions. Can you?

The doctor thought about this answer, then said, "Even four years may be too long. If we were willing to stimulate four regions initially, could we reawaken the whole brain in one year?"

Ecco shook his head.

? 2. Can you see why this is impossible?

"However, I can see how to do it in two years," he volunteered.

? 3. What four regions should be stimulated initially to wake up Baskerhound's entire brain in two years?

After the doctor had left, I still had a question for Ecco. "Tell me," I said, "what was the relationship between Baskerhound and the Oddists?"

"Though they never told us, the Oddists are radical environmental libertarians. Baskerhound served as their military arm," he replied.

"What exactly do they believe?" Evangeline asked.

"Let me read from the pamphlet Fujisawa gave me," said Ecco:

> Like the Marxists who thought that capitalist societies would fall in their hands like overripe fruit, the classical libertarians assume that governments will eventually collapse under their own weight. But that is not so. Behold! State control in the European or North American model is outwardly healthy, while inwardly sick.
>
> We Oddists find the so-called free societies of North America and Australia too constricting. We shall create a libertarian society from scratch by any means necessary. Antarctica will be our base. From there we will go forth to save the world from the state-bound destroyers of nature.

"But why did they pose all those questions to us?" I asked. "Were they really working out new rituals?"

Ecco shook his head. "The Oddists gave me a test to prove who I was. I passed. They could not afford to give Baskerhound's letter to the wrong man."

45. Three Notes

At that moment, President Yerrek called. Ecco put him on the speaker-phone.

"Dr. Ecco," the president said, "I wanted to thank you and your friends personally. You have saved our country for a second time. Is there anything I can do for you?"

"Give my cousin Michael, a Chinese dissident, asylum," said Evangeline.

"Done," said the President. "Anything else?"

"No, sir," said Ecco. "We ask nothing more than to retreat to obscurity."

"Very well, Dr. Ecco, we will award you no medals," said the president with a chuckle. "Oh, I forgot to tell you. The Director has escaped from the hospital where he was recuperating. A coded note in his handwriting has been found in Skagway, Alaska. A messenger will bring it to you within an hour. Send back the decoding with him, won't you?"

"Yes, of course," said Ecco.

The messenger arrived and Ecco and Evangeline studied the note.

Geefy efle lze Yecef. Feddeo ee lzeje efv I'dd te yefe.

After they decoded the message, Ecco wrote out the cleartext and gave it to the messenger. Then he opened a package of oatmeal raisin cookies. He wrapped a cookie in a slice of Munster cheese and ate it.

"By the way Jacob," Evangeline said as she too reached for a cookie, "there is a small incongruity you haven't yet resolved."

"Oh, you mean about the code?" Ecco asked.

"What are you talking about?" I asked.

"Except for the message from Jacob the encoded messages we have received since you and Jacob saw the Oddists have all used the same code," Evangeline said.

"You mean Baskerhound, Admiral Trober, and the Director all used the same code?" I asked.

"That's right, Scarlet," said Ecco as he handed me a cookie. "And we have no idea how that is possible. Any theories?"

Optional Contest Puzzle: What is your theory? Use abduction.

Solutions

The Coded Letter

Graffito ergo sum.
Scrawling on a building near
Washington Square Park

1. Letter

My Dear Evangeline and Scarlet,

Your attempts to pick up my trail have been diligent, though according to my captor laughably incompetent. Please don't take offense, for he is an intolerable snob.

Because he is using me to solve many difficult puzzles a day, he gives me privacy and time to think. The result is that I can write to you at my leisure. I have my doubts that this letter will reach you, but I have a plan to leave it in a grocer's stockroom.

Oh, yes, my captor is Benjamin Baskerhound, the math prodigy who became a philosophy professor at Princeton, now turned kidnapper. He may enter other careers very soon. With the long-winded introduction out of the way, you will find it easy to decode the rest in spite of the fact that the code is rotating by one letter right now.

Baskerhound and his staff are continually on the move. He controls his affairs by cellular telephone, using private satellite channels. He finds ready welcomes in lands whose people admire rebels. We are now in Uruguay, a land filled with exiled Basques who had fought against Franco in Spain. Before this, we were in Hopi country, where the traditional suspicion towards outsiders was dispelled when Baskerhound (and I) helped to locate a giant aquifer. The code is rotating again. We stay in a place for a day, a week, a month. Then we are off, usually in Baskerhound's LearJets.

Baskerhound is so confident of his ability to evade detection that he allows me to play the tourist—accompanied by my "bodyguards," of course. The code is rotating.

Look for me in Uruguay. I will try to figure out our next destination and will leave word with a cheerful and utterly naive horsebreeder on a dirt road near a village called Punta Ballena. Ask for the pirate's house and you won't be far away. When you find the horse-breeder, ask him whether he is racing any lame horses this year. Code changes again. I will not be able to use this code again.

The encoding of the message I leave next time will rotate once each time the message contains the letter . . . well, you will figure it out. The time after that, the code will be the same one that this letter began with, except that the message will be split into groups of eight characters. These groups will be written in random order, but some of them overlap.

Baskerhound will be quickly aware of any major police operation, so please be subtle. Let someone know where you are going in case you disappear.

Yours,
Jacob

Further Reading: Readers who enjoy breaking codes might think of subscribing to the *Cryptologia Journal* for some wonderful coding ideas.

Armada at Moher

2. Running Bulls

1) There is no way to have a race of four bulls on four tracks where even one bull can end on a different track and still have the routes meet only at the intersections. To see this, suppose that X-X′ is the leftmost track from which some bull, call it Toro, takes a cross-track towards the right. (Some bull must travel to the right.)

Number the cross-tracks starting at 1 with the uppermost cross-track. Suppose the cross-track Toro uses is cross-track i. But then no other bull can run from cross-track i and to cross-track $i + 1$ on track X-X′. So, there would be only three tracks to use between those two cross-tracks. That is too few tracks for four bulls who must never meet.

2) With five tracks, but only four bulls, four cross-tracks are enough. First, the permutation with at least one straight route (that is, one that uses no cross-tracks) reduces the problem to one of three bulls on four tracks. This is handled easily. If there are no straight routes, then there are two fundamentally different possibilities: two exchanges (two permutations of two tracks each) or zero exchanges. Here are representative hard cases: A-C′, B-D′, C-A′, D-B′ and A-C′, B-D′, D-A′, C-B′ (see figures on this page and on top of p. 198).

The method used was quite simple and general. Number the bulls 1, 2, 3, 4, where bull 1 will end at A′, bull 2 at B′, and so on. At the first cross-track, bull 4 goes to track E-E′, where it stays until the last cross-track. On the first cross-track, the bulls to the left of bull 4 each move right by one track. So, A-A′ is empty between the first and second cross-tracks.

At the second cross-track, bull 1 goes to track A-A′, where it remains until the end. Except for bull 4, any bulls to the right of bull 1 move one track to the left along the second cross-track. So, only bulls 2 and 3 may be out of order, and D-D′ is empty.

At the third cross-track, bull 3 goes to track D-D′. The bulls are now in the correct order. The fourth cross-track can now put them all on the correct tracks.

3) The permutation A-D′, B-A′, C-B′, D-C′ requires only two cross-tracks (see bottom figure on p. 198).

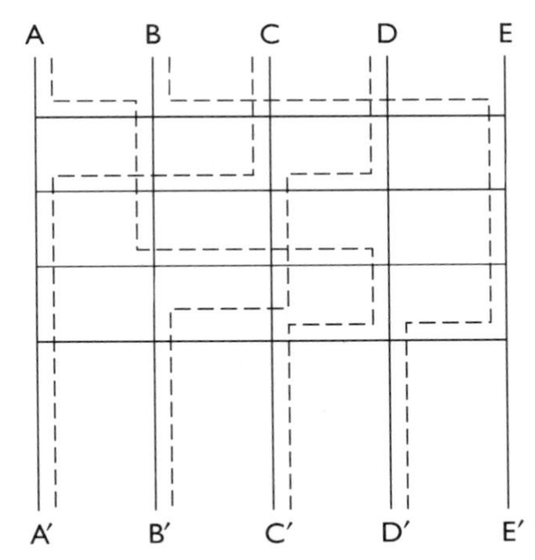

Permutation A-C′, B-D′, C-A′, D-B′.

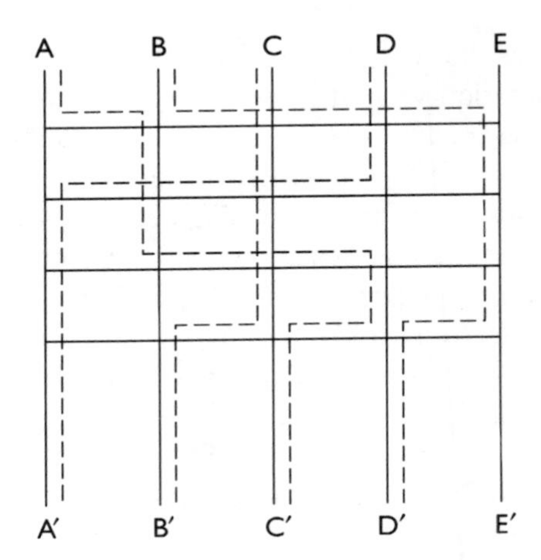

Permutation A-C′, B-D′, D-A′, C-B′.

Further Reading: This puzzle is inspired by the problem of routing wires on electronic substrates. Each layer of metal goes in a direction perpendicular to the layer above it, and the problem is to connect certain sets of locations. A readable introduction to the

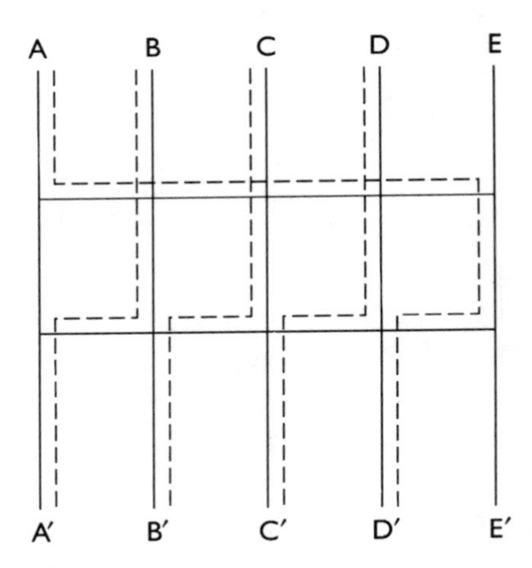

A solution requiring only two cross-tracks.

subject can be found in Jeffrey D. Ullman's *Computational Aspects of VLSI* (New York: Computer Science Press, 1984).

3. Humpty Dumpty at the Irish Cliffs

The code rotates each time an "h" is encountered. Here is the letter in decoded form:

Dear E and S:

It is early Saturday morning and I might just be able to deliver the letter to the good horse-breeder. We are on our way to Ireland. Baskerhound has cut short our stay. I'm not sure why, except that he made this decision after a short conversation with Donaldo Rumtopo, owner of El Casino. Perhaps he knows that you are coming, but I doubt it.

He seems suspiciously happy. "The Cliffs of Moher," he says rubbing his hands with pleasure, "the Cliffs of Moher. You will like them, Ecco, I am sure."

I am not. But follow me to Ireland. My next letter will be in the Kames museum in Cork. I have made a habit of visiting it whenever I'm in the area. You will find the curator to be a strange but sincere lady named Anne O'Connell. Local people think she's crazy with all her talk of Inca stone carvings in Ireland. But she is an outstanding genealogist. Tell her you are trying to find your cousin Issac O'Connell. She will give you the note.

J

4. El Casino

Both the solutions in both variants of the game are guided by the odds of obtaining each total with a pair of fair dice. The odds of getting a 2 are 1 in 36. The odds of getting a 3 are 2 in 36, since there are two combinations that can lead to a 3.

Here is a summary of the calculations:

TOTAL	ODDS OF GETTING THAT TOTAL
2	1 in 36
3	2 in 36
4	3 in 36
5	4 in 36
6	5 in 36
7	6 in 36
8	5 in 36
9	4 in 36
10	3 in 36
11	2 in 36
12	1 in 36

1) So, in the first game, the casino will win 21 out of 36 times. The roller will win 15 out of 36 times. If each bet is 1 dollar, over 36 rolls the roller will win 18.75 dollars, but the casino will win 21 dollars. That means that in each roll, the roller will lose an average of (21 − 18.75)/36 dollars, or 6.25 cents per roll when a dollar is bet.

2) In the second game, the analysis is more tricky. One way to look at the game is to say that the roller can be in one of two states: slow or fast. Initially, the roller is in the slow state.

In the slow state, he has a 21 in 36 chance of remaining in that state and losing, and a 15 in 36 chance of winning and moving to the fast state.

From the fast state, he has a 15 in 36 chance to win and stay, a 15 in 36 chance to lose and go to the slow state, and a 6 in 36 chance of staying where he is without any exchange of money. Like Evangeline's interlocutor, we ignore the last possibility since nothing changes as a result of a roll of 7. So from the fast state the roller has the same probability of winning or losing per significant roll. The figure on page 201 shows the transitions.

Let p_slow be the probability that the roller is in the slow state and p_fast the probability that he is in the fast state. Clearly, p_fast = 1 − p_slow. More interesting, the probability that a roller

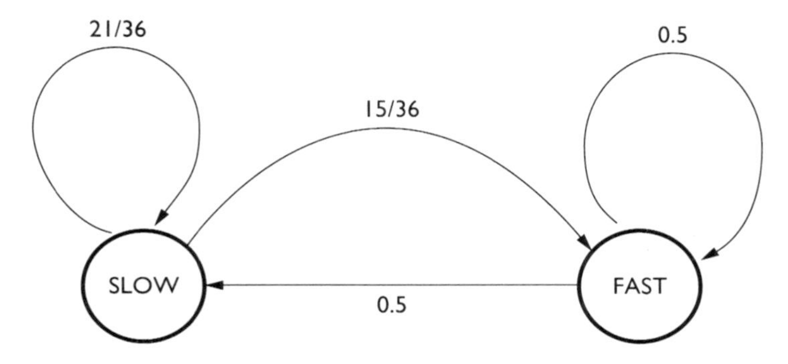

Effective transition likelihoods from the slow state to the fast state and back.

will be in p_slow after the next significant roll is (21/36)p_slow + (0.5)p_fast. Solving these two equations for these two unknowns gives the likelihood of being in the slow state as 6/11 and of being in the fast state (after a winning roll as opposed to just a draw) of 5/11.

Now, from state slow, the roller's expected win, win_slow (per dollar bet) will be

$$(15/36)1.05 - (21/36) = -14.58 \text{ cents}$$

From the fast state, the roller's expected win per significant roll on a dollar bet, win_fast, is

$$(0.5)(1.25) - (0.5) = 12.50 \text{ cents}$$

Multiplying these expected wins by the probability of the states in which they occur, we get

$$\text{p\_slow} \times \text{win\_slow} + \text{p\_fast} \times \text{win\_fast} = -2.3 \text{ cents per significant roll}$$

3) In the proposed game, the probabilities p_slow and p_fast remain the same. What changes is win_slow and win_fast.

$$\text{win\_slow} = (15/36)1.25 - (21/36) = -6.25 \text{ cents per roll}$$

$$\text{win\_fast} = (0.5)(1.05) - (0.5) = 2.5 \text{ cents per significant roll}$$

Multiplying these expected wins by the probability of the states in which they occur, we again (by numerical coincidence) get

$$p\_slow \times win\_slow + p\_fast \times win\_fast = -2.3 \text{ cents per significant roll}$$

5. Greed and Getaways

1) The inspector might hear 13 different answers before he would be sure of the routes. Suppose the two routes were A and B. There are 19 partially honest, different reports containing two routes and 2 more containing just one route each. However, notice that any given route, say X, is established after [X, Y], [X, Z], and [X, W], where W, Y, and Z are mutually distinct, even if one of the others is also true.

So, the inspector might hear about 9 routes associated with A but not including B. Then he might hear a report consisting of a singleton [A]. Further honest (or half-honest) reports that are different from these first 10 must include B. So, the next 3 would include B ([A, B] might be one of the pairs), enough to establish the pair. Therefore 13 different reports suffice to determine the exact routes.

2) How many reports are needed to prove that A and B are correct? At most, five. Three are needed to establish A. If possible, choose one that includes B. Once A is established, three reports including A or two reports not including A are enough to establish B.

Fewer than five reports are not enough in at least some cases. Suppose that the only distinct reports received are [A, B], [A, C], [A, D], [B, C], and [B, D]. These establish A and B, but deleting any one of these opens up other possibilities.

6. Museum Tour

Except for the very first characters, each character in the cleartext is represented by two characters in the ciphertext. Here is the actual cleartext. Don't let the hyphens deceive you.

Dear Evangeline and Scarlet,

What is Baskerhound up to? The night we arrived in Moher, a U.S. Navy ship disappeared. So did five of Baskerhound's retinue--all ex-naval officers. Three more disappeared near Adare.

I don't know what they did, but Baskerhound was in an uncharacteristically good mood today at breakfast. "Ecco," he said, "the economic need for your services has passed. But don't pack your bags yet. We will be companions for some time to come. Maybe your two friends can be persuaded to join us."

I'm afraid of his good moods and am worried about your safety, my friends. You may be better off returning home. Maybe you can convince the Director to give you some protection. We are leaving today for Asia, I think.

Jacob

The Genius of Georgetown

7. Musical Messages

1) Here is a 14-bit encoding of those nine messages: 01000011011100.

2) Here is a sequence that encodes all 16 possible messages in 19 bits: 0000111101100101000.

3) Ecco was able to show a stronger result than requested: if he is given any 15 4-bit messages, then he can encode them in an 18-bit sequence. Demonstrate this by considering 16 ways to encode all 16 messages:

000111101100101000 0
0011110110010100001
0111101100101000011
1111011001010000111

1110110010100001111
1101100101000011110
1011001010000111101
0110010100001111011
1100101000011110110
1001010000111101100
0010100001111011001
0101000011110110010
1010000111101100101
0100001111011001010
1000011110110010100
0000111101100101000

As you can see, every possible message is the first four bits of one of the sequences. Suppose that some set of 15 4-bit messages does not include a given message, m. Just choose the 19-bit sequence from this list that has m as its first four bits and remove the first bit from that sequence. Clearly all 15 possible other 4-bit messages are in the truncated sequence.

4) To see a set of 14 4-bit messages that cannot be encoded in a 17-bit sequence, consider the set consisting of all messages except 1000 and 0001. The 14 messages in that set can fit in a 17-bit sequence only if every bit starting with the fourth one of the sequence terminates a message in the set. No messages outside the set are allowed.

Since 0000 is one of the 14 messages, it must be present in the 17-bit sequence. Since the sequence can't be all 0's, 0000 must be either preceded or followed by a 1. But that would be a message outside the set. Therefore any such sequence encoding all messages except 1000 and 0001 must contain at least 18 bits.

8. Elves Flip

1) Imagine a graph of the net number of heads (that is, the number of heads less the number of tails) on the vertical axis versus the number of flips on the horizontal access. After the first toss, the net number is either +1 or −1. Each toss changes the net number by +1 or −1. At the end, the net number is 3 (since there are seven heads and four tails).

Consider a path, call it T, on this graph that starts at +1, goes to zero at some time, but still ends at +3 after 11 flips (see the top figure on this page for an example). Now, construct a unique path T′ that starts at −1 and ends at +3 as follows: until T returns to the zero line, T′ is the mirror image of T (see the bottom figure on this page). After that T and T′ are the same. Every such T corresponds to a different T′ constructed in this way.

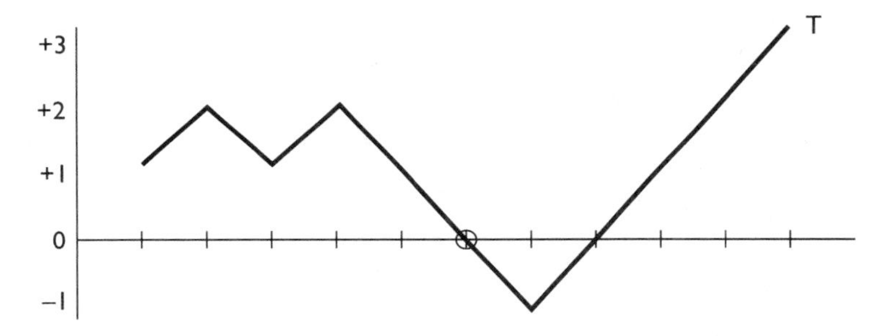

Example of a path that starts at +1 and ends at +3. Notice that this path dips below the x-axis, indicating that the bettor is losing some of the time.

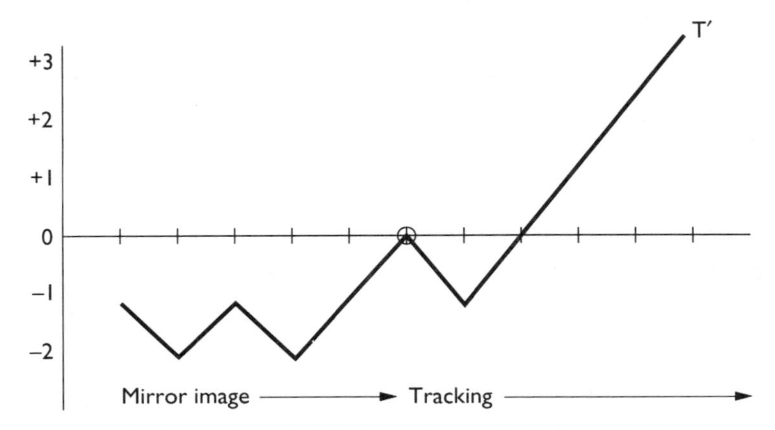

Another path that ends at +3, but starts at − 1. It is critical to observe that any path that starts at +1 and later hits 0 has a mirror image in a path that starts at − 1 and hits 0. After a mirroring path hits 0, its plot is the same as the original path.

So the number of paths that go first to +1, end at +3, and return to the zero line in between is exactly equal to the number that go first to −1 and end at +3. Of course, the latter all touch the zero line.

We want to compute the likelihood that heads ends at +3 and always leads. This is the same as the number of such paths divided by all possible paths induced by 11 coin flips. How many such paths are there? Well, there are all the paths that end at +3 minus those that end at +3 and touch the 0 net heads axis at least once after the first flip. The argument of the previous paragraph shows that the number of paths that go first to +1, touch the 0 net heads axis and end at +3 is equal to the number that go first to −1 and end at +3.

So the likelihood that a path ends at +3 and always leads = the likelihood that a path ends at +3 − 2 × likelihood that a path ends at +3 and goes first to −1.

This equation is

$$\binom{11}{7} \times 1/(2^{11}) - 2 \times (1/2) \times \binom{10}{7} \times 1/(2^{10})$$
$$= 330/2048 - 240/2048 = 90/2048 = 45/1024$$

2) Use the method of the first solution and Professor Scarlet's calculations to compute the result for each possible outcome. Then use Professor Scarlet's observation that the results from the different outcomes can be added together, because they are mutually exclusive. The arithmetic shows that the probability that the player will be ahead from the first to the last toss is 126/1024 or just under 1/8. So, the casino will make money if it pays 7 to 1 odds but only a little.

Further Reading: The reflection principle used in this solution makes many difficult probabilistic questions much easier. For other tricks on how to make probabilistic problems easy, see the classic book on probability ***An Introduction to Probability Theory and Its Applications*** by William Feller (New York: John Wiley & Sons, 1968).

9. A Problem of Protocol

Visitor 1 (chief Rukati) puts on all three gloves A, B, and C (with C on the outside) and shakes hands with the Secretary General. We might denote this as $1)_A)_B)_C$ SG.

Visitor 2 (assistant Rukati) puts B inside C and shakes hands with the Secretary General. Denote as $2)_B)_C$ SG.

Visitor 1 (chief Rukati) uses glove A to shake hands with the chief Tarak (Visitor 5). Denote as $1)_A$ 5.

Visitor 3 (interpreter) uses glove C to shake hands with the Secretary General. Denote as $3)_C$ SG. At the same time, Visitor 2 (assistant Rukati) uses glove B to shake hands with the assistant Tarak (Visitor 4). Denote as $2)_B$ 4.

Visitor 4 (assistant Tarak) puts glove C on top of reversed glove B to shake hands with the Secretary General. Denote as $4(_B)_C$ SG; the orientation of the parenthesis indicates that glove B was reversed.

Visitor 5 (chief Tarak) puts gloves B and C over reversed glove A to shake hands with the Secretary General. Denote as $5(_A)_B)_C$ SG.

10. Tropical Antarctica

Was your design much different from the drawing on p. 208?

11. 100-Day Rockets

1) The basic idea behind the 100-hour solution is very simple. The 100 trucks work in synchrony. Each truck goes to 10 sites. At each site, it takes 10 component copies produced from that site. Thus, the truck has 10 copies of each of 10 components. Then, the truck deposits a copy of all 10 components at each of 10 destination sites. If 100 trucks do this, then only one round of this is necessary. We need only one loading dock per factory.

Here is a more explicit description. We have 100 trucks and 100 sites, each numbered 1 to 100. Two pieces of terminology:

Let x div y be the quotient of x divided by y.

Let x mod y be the remainder.

For example, 74 div 10 is 7, and 74 mod 10 is 4.

This is what truck i does. Let $k = (i - 1)$ div 10. Let $j = i$ mod 10.

Truck i begins at $1 + 10k + j$, collecting 10 component copies from there, then collects 10 component copies from $1 + 10k +$

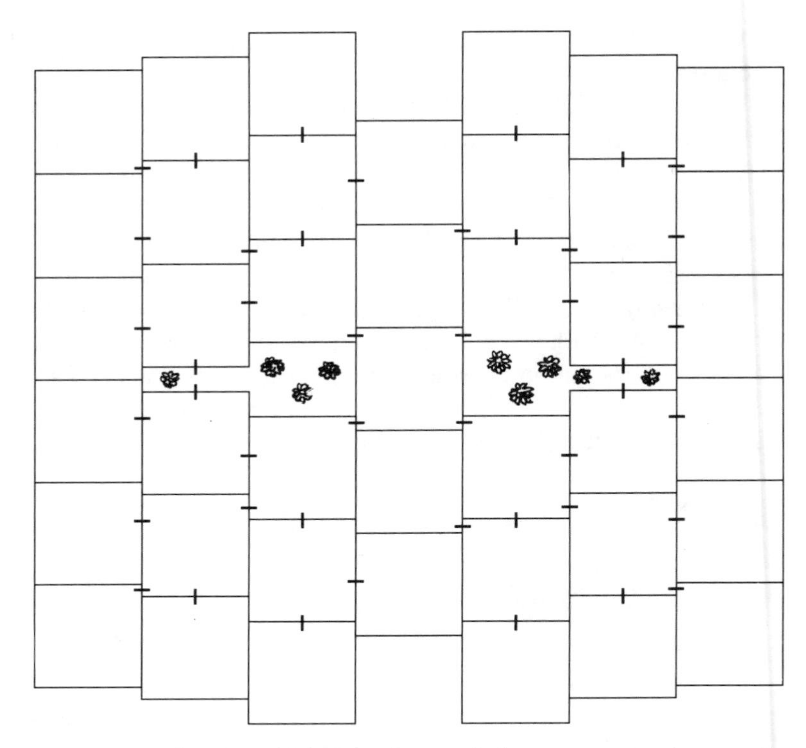

A 41-room design for which there is a path from any room to any other going through six or fewer doors. The circular figures indicate plantings in two interior gardens. Four rooms have sealed windows overlooking the gardens.

$((j + 1) \bmod 10)$, and so on up to and including $1 + 10k + ((j + 9) \bmod 10)$. In the second phase, truck i begins at $1 + 10j + k$ (notice the reversal of j and k), deposits one copy of each of the 10 components it has there, then does the same at $1 + 10j + ((k + 1) \bmod 10)$ and so on up to $1 + 10j + ((k + 9) \bmod 10)$.

If you find this too symbolic, try to follow it with trucks 1 through 10, and 20, 30, 40, 50, 60, 70, 80, 90, and 100. You'll see that sites 1 through 10 get all the 100 components.

2) It is possible to save money and time by making the last pickup site be the first destination site.

Let $k = i$ div 10.

Let $j = i \bmod 10$.

Truck i begins at $1 + 10k + j$, collecting 10 component copies from there, then collects 10 component copies from $1 + 10k + ((j + 1) \bmod 10)$, and so on up to and including $1 + 10k + ((j + 8) \bmod 10)$. At $1 + 10k + ((j + 9) \bmod 10)$, truck i delivers the 9 different components it has collected so far, then picks up 9 components from that site. In the second phase, truck i deposits one copy of each of the 10 components it has at $1 + 10\,((k + 1) \bmod 10) + j$, $1 + 10\,((k + 2) \bmod 10) + j$, and so on up to $1 + 10\,((k + 9) \bmod 10) + j$, thus avoiding a visit to $1 + 10k + j$. So, the total number of factories visited is 19 instead of 20.

3) To finish the job in 40 hours requires finishing it in eight trips. That is, eight visits to factories (since the first visit to a factory requires a trip from the truck depot). Here is the best method Dr. Ecco could think of. Use 1500 trucks. Each factory needs to use only 15 loading docks. In the first trip, send 15 trucks to each factory.

Consider the trucks at factory k. Number them 1 to 15. This is what truck i at factory k does.

If $i < 10$, take 7 components from site k and deliver them to sites $1 + ((k + 1 + 7 \times i) \bmod 100)$, $1 + ((k + 2 + 7 \times i) \bmod 100)$, $1 + ((k + 3 + 7 \times i) \bmod 100)$, and so on up to $1 + ((k + 7 + 7 \times i) \bmod 100)$.

If $14 > i \geq 10$, take 6 components from site k and deliver them to sites $1 + ((k + 1 + 10 \times 7 + 6 \times (i - 10)) \bmod 100)$, $1 + ((k + 2 + 10 \times 7 + 6 \times (i - 10)) \bmod 100)$, $1 + ((k + 3 + 10 \times 7 + 6 \times (i - 10)) \bmod 100)$, and so on up to $1 + ((k + 6 + 10 \times 7 + 6 \times (i - 10)) \bmod 100)$.

If $i = 14$, then deliver to $1 + ((k + 95) \bmod 100)$, $1 + ((k + 96) \bmod 100)$, $1 + ((k + 97) \bmod 100)$, $1 + ((k + 98) \bmod 100)$, skip sending to self and then $1 + ((k) \bmod 100)$, and $1 + ((k + 1) \bmod 100)$.

If $i = 15$, then deliver to $1 + ((k + 2) \bmod 100)$, $1 + ((k + 3) \bmod 100)$, $1 + ((k + 4) \bmod 100)$, $1 + ((k + 5) \bmod 100)$, $1 + ((k + 6) \bmod 100)$, and $1 + ((k + 7) \bmod 100)$.

So, 99 copies of component k are removed from that factory and delivered to all the other sites.

Further Reading: This problem is inspired by parallel algorithms for matrix transposition. In that problem, all the data from a single row are spread among all the columns. Here we are spreading all the components from a single factory among all the factories. There are many conferences and journals on the subject of parallel algorithms. A particularly good series is entitled Symposia on Parallel Algorithms and Architectures and is sponsored by the Association for Computing Machinery (New York).

12. Three-Finger Shooting

1) To confirm Scarlet's assertion, merely observe that since each player has three choices, there are 3 × 3, or 9, possibilities. For the choices given, five of the possibilities give an even sum and four give an odd sum.

2) The player always chooses two fingers. She will win whenever the casino plays one or three, which will happen two out of three times.

3) If the casino announces a strategy in which it chooses two fingers less than half the time, then the player can win on the average by always putting out two fingers. If the casino announces a strategy in which it chooses two fingers more than half the time, then the player can win by never putting out two fingers. If the casino puts out two fingers exactly half the time, then the player can use any strategy at all and the game will be even. Thus, the casino cannot announce a strategy and expect to win.

4) If neither side announces a strategy, each side can guarantee a draw by choosing two fingers half the time and one or three the other half. If either side uses some other strategy and reveals it, the other side can exploit it as in the previous solution. For example, if the casino chooses two fingers more than half the time, the player should choose two fingers less than half the time.

5) If the player chooses two fingers half the time and three fingers half the time, the casino has no strategy it can use to win. To see this, consider all the times the casino chooses one. On the average the player will choose two half of those times and lose and three half of those times and win. The same goes for those occasions when the

casino chooses three, except that the player wins on two and loses on three in that instance. Surprisingly, the player always wins when the casino plays two. So, the casino's best response is to choose one finger half the time and three fingers half the time.

Justifying the Means

13. Sand Magic

1) Ask the Sand Counter to leave the stage. Remove a small quantity of sand from the bucket and count the grains. Smooth the remaining grains on the surface. When the Sand Counter returns, ask him to determine how many grains you have removed. His task hasn't become much easier, nor does he tell you anything that you don't already know. You can repeat the test as many times as you like to ensure that he is not simply lucky.

2) Here is Ecco's explanation: "As an impersonator, I would claim to a third party that I was an Amazing Sand Counter. The third party would put me to the test. When he showed me the first bucket, I would start a test on the true Sand Counter and show him the same bucket. When the third party showed me the bucket with some sand removed, I would do the same to the Amazing Sand Counter. The Sand Counter would tell me how much sand was removed. I would give the third party the same number and the third party would be convinced!"

Further Reading: In computer systems, protocols similar to the one Ecco used to test the Amazing Sand Counter are used to verify the identity of computerized agents. An agent A proves its identity by answering questions about some code (analogous to the Sand Counter's professed skill) that only A can know. However, A does this without divulging the code, to prevent A's interlocutor from masquerading as A. For this reason, these are called zero-knowledge proofs. Nice introductory discussions of zero-knowledge proofs can be found in *The Mathematical Tourist: Snapshots of Modern Mathe-*

matics by Ivars Peterson (W. H. Freeman and Co., 1988). The second puzzle shows that even if A's interlocutor, call it B, gains no knowledge, B may still convince some other agent C that B is C. B basically uses a strategy of forwarding (or replaying) to C the messages that B receives from A and then returns the messages from C back to A. In the jargon, the imposter B would have engaged in a "replay attack."

Further Thinking: (suggested by Professor Michael Rabin of Harvard): Suppose you prove that the Amazing Sand Counter has the power he claims to have. Now, you fill a bucket with sand and ask him how many grains it has. Suppose he tells you a number. Can you design an experiment to test whether he is telling the truth?

14. Signals and Echoes

1) Since B is the most frequent message symbol, use the shortest possible encoding for it. Thus,

To convey B, send E.

To convey A, send AA.

To convey C, send CC.

To convey D, send AC.

To convey E, send CA.

Since A, C, and D all have about the same frequency, using a one-tone encoding for one of them will force one of the others to be of length three, so nothing would be gained. In that case, the encoding of E would be one tone longer, so the encoding would be slightly less efficient. Using E as part of the encoding of A, C, D, or E will not help and may lead to confusion.

2) Since an echo can be confused with the same tone repeated, our encoding system will not use repeated tones. Also, since we cannot use pauses, we must ensure that repeats or echoes do not lead us to believe that two symbols are being sent when in fact only one is sent. So, we will use E as a terminating tone (acting as a period in a sentence). Here is the encoding:

To convey B, send AE.

To convey C, send CE.

To convey A, send ACE.

To convey D, send CAE.

To convey E, send ACAE.

Further Reading: The figure that accompanies this puzzle in the text is called a confusion graph. For more information on this and other strange graphs, see ***Graph Theory and Its Applications to Problems of Society*** by Fred S. Roberts (Philadelphia: Society for Industrial and Applied Mathematics, 1978).

15. Territory Game

1) If the British take the center, then the Argentines can take site 5. If the British then take any site to the left of 12, the Argentines should take the site just to the right of the rightmost British site, securing at least seven sites. If the British take 12 or 13, then the Argentines should take position 8, allowing them to control at least seven sites.

2) If the British take sites 4 and 10, they are sure to win. Consider the following three cases (the others are clear losers for Argentina or are symmetric to the three cases presented here):

Case 1: The Argentines take 3 and 11. Then the British have sites 4–10, or seven sites.

Case 2: Argentina takes site 3 and any site between 5 and 9. Then the British secure sites 10–13 and at least as many sites

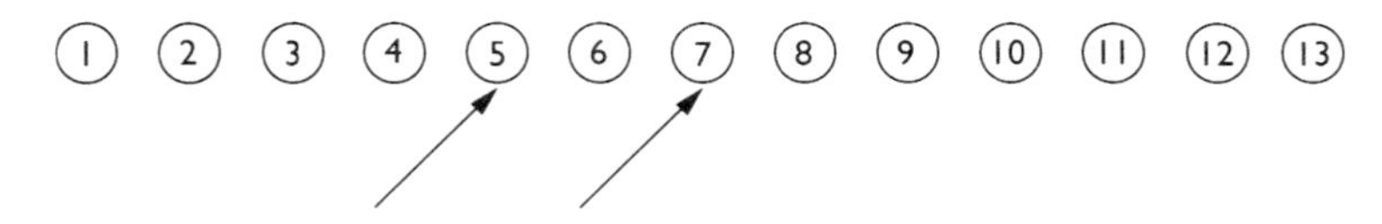

If the British take site 7, then the Argentines will win by choosing 5.

between 4 and 9 as the Argentines. So, the British win 7 to 6 or 6 to 5.

Case 3: Argentina takes 5 and 9. Then the British control eight sites.

3) If the British take site 4, then the Argentines must take site 10 (by the solution to 2). After that the British take 11. Unless the Argentines play symmetrically (taking position 3), they will lose. Playing symmetrically guarantees a tie.

4) The British can never force a win in two moves. If their first move is anything other than 4 or 10, they can do no better than a tie. (Before seeing the proof of that, note that if the British first move is 4 or 10, then the Argentines play symmetrically.) If the first British move is one of sites 1–7, but not site 4, then the Argentines take site 4. (They play symmetrically for the higher numbered sites.) The second British move must be 10, since we have shown that if either side has 4 and 10 it will win. So, there are two cases: If the first move of the British is 5, 6, or 7, then the Argentines take site 11 and control seven sites. If the first British move is 1, 2, or 3, then the Argentines take site 11 and control at least six sites, with one site uncontrolled by either side.

Further Thinking: The really ambitious reader may try to find a general winning strategy for the Territory Game in one dimension or even two. The two-dimensional version makes for a nice video game. In fact, there is a further variant in which one plays the game for k moves and then plays "snipe" for m moves where $m < k$ (a move of snipe consists of removing an opponent's choice). So, at the end, each player has taken only $m - k$ sites.

16. Drugs and Interdiction

1) If the bases are numbered I–VII, then put two jets at I and at VII and one at all other bases. If a single intruder approaches the sections controlled by I or VII, fly no reinforcements from neighbors, unless a second intruder approaches. If a single intruder approaches II–VI, fly a single jet from both neighbors of the base being approached. Three jets are enough to face both intruders. If the second intruder ap-

proaches one of the now-empty sections, the jet from the corresponding base would fly back as well as a jet from its neighbor.

2) This is not the only solution with nine jets, but nine is clearly the minimum possible, since three jets must be within eight minutes of each base's section. So, I and II must have three jets between them. Similarly for VI and VII, similarly for III, IV, and V.

Jungle Killers

17. Amazon Exchange

2A → N → 2G → 4C 4D → N H C → H C 2G → H C 4C 4D → H 2C N H → 2H 2C 2G → 2H 2C 4C 4D → 3H 3C N → 3H 3C 2G → M 2H C 2G → M 2H A 2G → M 2H B C F 2G → M 2H B A F 2G → M 3H 3G B F → Hale plus a few odds and ends.

Further Reading: This puzzle was inspired by general rewriting systems known as Thue systems after the Norwegian mathematician who invented them. Two excellent texts that discuss Thue systems are *Computability, Complexity and Languages* by Martin D. Davis and Elaine J. Weyuker (New York: Academic Press, 1983) and *Elements of the Theory of Computation* by Harry Lewis and Christos Papadimitriou (Englewood Cliffs: Prentice-Hall, 1981).

18. Mutual Admiration

Evangeline and Scarlet conclude that 9 and 10 must be good. Here is their reasoning: Because 18 and 19 accuse one another of being bad, at least one of them must be bad. Therefore all of 11 – 17 and 20 are bad, since they all claim that both 18 and 19 are good. Since 18 and 19 both say that 11 – 17 and 20 are good, 18 and 19 must, in fact, both be bad. Since 1 – 5 claim that 14 is good, 1 – 5 must be bad. Since 6 – 8 claim that 3 is good, 6 – 8 must be bad. 9 – 10 could be good and 21 – 25 could be bad, however the opposite is not possible. So, if anyone is good, 9 – 10 must be good. Further, 21 – 25 could be good. Everyone else (including the chief) is bad for sure.

Double Escape

19. The Octopelago Problem

1) Alternate the direction in which fares are collected. So, either all routes *entering* a particular island require paying a fare or all routes *leaving* the island require paying a fare. In the figure on this page, an arrow indicates the direction a fare is collected (the return ferry ride is free). Each fare should be $2.00.

2) If the islands are numbered 1 to 8, put a ferry between 1 and 4 and between 2 and 7. At every island either all ferries leaving that island should require a fare of $2.00 or none should. Every round-trip journey requires an even number of ferry rides.

20. MicroAir

1) Let us number the cities from 1 to 7. Here are the routes of the seven planes:

first plane 1 2 4 1 2 4 . . .

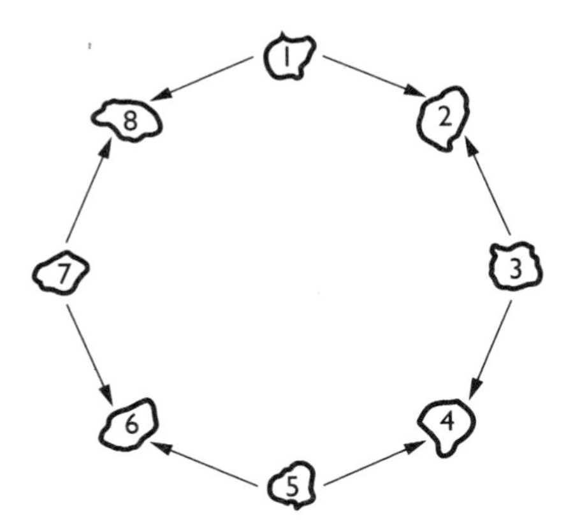

The young mathematician's solution: the arrow indicates the direction in which to collect fares, so all passengers pay as much for any round-trip journey as they would have with one-way fares.

second plane 3 4 6 3 4 6 . . .

third plane 2 3 5 2 3 5 . . .

fourth plane 4 5 7 4 5 7 . . .

fifth plane 1 5 6 1 5 6 . . .

sixth plane 2 6 7 2 6 7 . . .

seventh plane 1 3 7 1 3 7 . . .

2) If there are only six planes, then we can guarantee arrival within 5 hours and 20 minutes by the following schedule. Notice that each plane begins its cycle again 7 hours after it starts:

first plane starts at 0:00, 1 2 3 4 5 6 7 1 2 3 4 5 6 7 . . .

second plane starts at 2:20, 1 2 3 4 5 6 7 1 2 3 4 5 6 7 . . .

third plane starts at 4:40, 1 2 3 4 5 6 7 1 2 3 4 5 6 7 . . .

fourth plane starts at 1:10, 7 6 5 4 3 2 1 7 6 5 4 3 2 1 . . .

fifth plane starts at 3:30, 7 6 5 4 3 2 1 7 6 5 4 3 2 1 . . .

sixth plane starts at 5:50, 7 6 5 4 3 2 1 7 6 5 4 3 2 1 . . .

Further Reading: This puzzle takes its inspiration from "combinatorial design" problems. A clear introduction to the field may be found in A. P. Street and D. J. Street's book *Combinatorics of Experimental Design* (New York: Oxford University Press, 1987).

21. Personals

1) Lk wrl ab xmm cbscnt cp Iddjezyn xej Kyof.

2) my apartment is dusty. see you soon. ecco.

22. Shark Labyrinth

This is how Ecco explained it: "I start at A with the chum—the fish from dinner—and swim towards B. I leave the chum at B. I then kill the shark at C. This attracts all the sharks except those at G and H. I return to B and then leave the chum 10 meters from B towards C. I wait and kill the first shark that approaches. That will leave me time to swim from B to J to F."

23. A Question of Inheritance

1) Five weighings are sufficient.

Give coins the labels A, B, C, D, E, F, G, H, I, J. As a shorthand, allow "weigh XYZ" to mean weigh X, Y, and Z together. Also allow "one of X, Y" to mean "either X or Y."

```
weigh ABC
 if 2 of ABC are fake, then weigh A first and B second. Done.
 if 1 of ABC is fake, then
  weigh ADE
   if 2 of ADE are fake (A is fake and one of D and E is fake), then weigh D.
Done.
   if 1 of ADE is fake, then
    weigh AFG
     if 2 are fake (A is fake and one of F and G is fake), then weigh F. Done.
     if 1 is fake (A is the only fake in A-G), then
      weigh HI
       if 1 is fake then weigh H else weigh J. Done.
     if 0 are fake (one of B,C is fake and one of D,E is fake), then
      weigh B first and D second. Done.
   if 0 of ADE is fake (ADE are good, one of B,C is fake), then
    weigh BFG
     if 2 are fake (B must be fake), then weigh F. Done.
     if 1 is fake (B is fake or both C is fake and one of F,G is fake)
      weigh BHI
       if 2 are fake (B and one of H,I is fake), then weigh H. Done.
       if 1 is fake (B alone is fake among A-I), then weigh J. Done.
       if 0 is fake (C and one of F,G is fake), then weigh F. Done.
     if 0 is fake (C is only fake in A-G), then
      weigh HI
       if 1 is fake then weigh H else weigh J. Done.
 if 0 of ABC is fake, then
  weigh DEF
   if 2 of DEF are fake, then weigh D first and E second. Done.
   if 1 of DEF is fake, then
```

```
    weigh DGH
      if 2 are fake (D is fake and one of G,H), then weigh G. Done.
      if 1 is fake (D or both one of E,F and one of G,H), then
        weigh EGI
          if 2 are fake, then E and G are the fakes. Done.
          if 1 is a fake, then
            weigh EFG
              if 2 are fake, then F and G are the fakes. Done.
              if 1 is fake, then E and H are the fakes. Done.
              if 0 is fake, then D and I are the fakes. Done.
          if 0 is a fake (D or both F and H are fakes), then
            weigh DJ
              if 2 are fake, then D and J are fakes. Done.
              if 1 is fake, then D alone. Done.
              if 0 is fake, then F and H are the fakes. Done.
      if 0 is fake (one of E,F is fake), then
        weigh IE
          if 2 are fake, then I and E are fakes. Done.
          if 1 is fake, then
            weigh JE
              if 2 are fake, then J and E are fakes. Done.
              if 1 is fake, then only E is fake. Done.
              if 0 is fake, then I and F are fake. Done.
          if 0 is fake (F is fake), then weigh J. Done.
  if 0 of DEF is fake (A-F are all good), then
    weigh GH
      if 2 are fake, then G and H are fakes. Done.
      if 1 is fake, then
        weigh GI
          if 2 are fake, then G and I are fakes. Done.
          if 1 is fake (G is fake or both H and I are fake), then
            weigh GJ
              if 2 then G and J are fakes. Done.
              if 1, then G alone. Done.
              if 0, then H and I. Done.
          if 0 is fake (H is the only fake in A-I), then weigh J. Done.
      if 0 is fake (A-H are all good), then weigh I first and J second. Done.
```

Four weighings are too few, assuming only three coins can be weighed at a time. To see that, notice that there are $\binom{10}{2}$ possible distinct pairs of fake coins, that is, 45. If there is only one fake, there are an additional 10 possibilities. There remains the possibility of no fakes. So there are 56 possible configurations altogether. Suppose that the first weighing gives the answer zero. Then, at best, three coins have been eliminated, leaving $\binom{7}{2}$ possible configurations with two coins, that is, 21. Add to that 7 with one coin and 1 with zero coins and you have 29 possibilities all together. Each subsequent weighing can give at most three answers, 0, 1, or 2. So, you cannot guarantee to reduce the number of possible configurations by more than a factor of three each time. Since there are only three weighings left, you can only reduce the number of configurations by a factor of $3 \times 3 \times 3$, or 27. That is not enough.

2) For the case where you might weigh more coins, five is still the minimum number of weighings. Let us assume that there is no ambiguity in the counts from the number of weighings. It is still the case that each weighing can give at most three answers, 0, 1, or 2, since there are only two fakes at most. So, the number of configurations cannot be guaranteed to be reduced by a factor of more than three. Clearly, weighing three or fewer coins in the first weighing will not work by the argument above. Suppose four are weighed in the first weighing with the result that one is fake. Then the number of possible configurations is the number within the group of four times the number within the remaining six. The number of possible ways in which one fake can be in a group of four is clearly 4. The number of ways zero or one fakes can be in a group of six is $1 + 6 = 7$. The product is 28. In three weighings, you can reduce the number of configurations only by a factor of 27, so three weighings are not enough.

Let us try the other cases. Similar reasoning works throughout. If five coins are weighed on the first weighing and one is bad, then there are $5 \times (5 + 1) = 30$ possibilities:

Case: six coins on first weighing and one bad. Then there are $6 \times (4 + 1) = 30$ possibilities.

Case: seven coins on first weighing and one bad. Then there are $7 \times (3 + 1) = 28$ possibilities.

Case: eight coins on first weighing and two bad. Then there are $\binom{8}{2} = 28$ possible configurations.

Case: nine coins on first weighing and two bad. Then there are $\binom{9}{2} = 36$ possible configurations.

Case: ten coins on first weighing and two bad. Then there are $\binom{10}{2} = 45$ possible configurations.

24. The Toxicologist's Puzzle

1) Step 1: Use filter 2. Whatever doesn't go through is some combination of {ABC, D, E}. What passes through is some combination of A, B, and C, with at least one of those constituents missing.

Step 2A: Use filter 1 on the part that passes through in step 1. What remains here is exactly one of {A, BC}. What passes through is exactly one of {B, C}.

Step 2B: Use filter 5 on the material that doesn't pass through in step 1. If anything goes through it is D. Whatever doesn't go through is some combination of {ABC, E}.

Step 3A: Take the material that doesn't pass through in step 2A and use filter 5. Either it all goes through or none of it goes through. If it all passes through, it is A. If none of it passes through, it is BC.

Step 3B: Take what goes through in step 2A and use filter 3. Either it all goes through or none of it goes through. If it all passes through it is B. If none of it passes through it is C.

Step 3C: Take what doesn't go through in step 2B and use filter 4. Whatever passes through is E. Whatever doesn't pass through is ABC.

2) One week is not enough because there is no way to distinguish ABD from ABDE in one pass through any or all of the membranes.

3) If only one constituent is present, just filters 1, 2, and 3 are needed and one week is enough.

4) Step 1: Use filter 1. Whatever passes through cannot contain both B and C. So, either some combination of B and D passes through or some combination of C and D, but not both.

Step 2A: Take what passes through filter 1 and pass half through filter 3 and half through filter 2. If any of it passes through filter 3, then B is present; whatever does not pass through is D and C is not present. If nothing passes through filter 3, then B is not present. So, what goes through filter 2 will be C and what does not will be D.

Step 2B: What does not pass through filter 1 is some combination of A and E. Pour that into filter 5. What passes through is A, what doesn't is E.

Further Reading: Readers interested in a lively introduction to nineteenth-century toxicology may wish to consult *The Century of the Detective* by Jürgen Thorwald (New York: Harcourt, Brace & World, 1965).

Battle for a Continent

25. How to Steal a Submarine

1) Call the imposter C. When C talks to B, C wants B to believe it is talking to A. Suppose there is a lot of traffic among A, B, and C. C waits until it is contacted by A.

i. A → C (broadcast): $S_A(P_C(\text{I am A}))$. C doesn't respond immediately to A, but instead replays an old broadcast message from A to B, at low enough power so A doesn't hear it.

ii. C → B (low-power broadcast): $S_A(P_B(\text{I am A}))$. Eventually, B responds to C. B responds as if it is talking to A.

iii. B → C (private): $S_B(n, P_A(\text{I am B}))$. C applies P_B to this message and thereby obtains n. Finally, C is ready to respond to A's request.

iv. C → A (private): $S_C(n, P_A(\text{I am C}))$. A thinks it is talking with C (and it actually is), so it responds obligingly with:

v. A → C (private): $P_C(S_A(n))$. Now, C can take this response and apply the following operations:

$$P_B(S_C(P_C(S_A(n)))) = P_B(S_A(n)).$$

This is what C needs to convince B that it (C) is A.

vi. C → B (private): $P_B(S_A(n))$.

So, C has successfully masqueraded as A as far as B is concerned.

2) All the attacks done so far have depended on varieties of replay attacks: that is, the imposter remembers a broadcast message, then rebroadcasts it. One way to counter such attacks is by using time-stamps. The new protocol requires just two messages. In each case let "clock" represent the time when the message is sent.

i. A sends to B via broadcast: B, $S_A(P_B(\text{clock,I am A}))$.

ii. B responds to A privately: $S_B(P_A(\text{clock,I am B}))$.

Since everyone shares a global clock, replays won't be believed unless they can exploit the small differences in the times the various ships receive the clock signals. The one-second delay to rebroadcast is important because a would-be imposter C can't exploit those time differences, since two ships will never be more than a fraction of a light-second apart.

Further Reading: This puzzle was inspired by an elegant paper, "*A Logic of Authentication*" by Michael Burrows, Martin Abadi, and Roger Needham, presented at the Association for Computing Machinery's 1989 Symposium on Operating Systems.

26. Missile Roulette

1) Ecco should choose the other card that Baskerhound leaves face down, even though he will have to win twice.

2) By choosing the other card that Baskerhound leaves face down, Ecco will win 12 out of 13 times. If he switches his choice both times, he will win both games with probability $(12/13)^2$, whose combined probability is greater than 11/13.

To understand the truth of the claim, think about this: Suppose that Ecco plays this game 13 times with Baskerhound. On the average, 12 out of those 13 times, Ecco will point initially at a card that is not an ace. In every one of those games, Ecco would win by choosing the other card that Baskerhound leaves face down. Only 1 in 13 times would Ecco win by sticking with his initial guess.

27. Finding the Target

Since D is west of B and because of the 3 – 4 – 5 ratio of the distances DB, BX, and DX, the three points DBX form a right triangle with X either due north or due south of B.

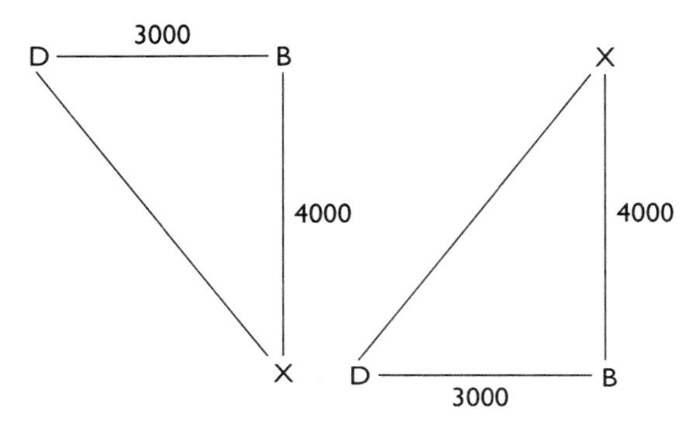

Two possible configurations of the triangle DBX, which must be a right triangle because of the 3-4-5 ratio among the sides. B is east of D.

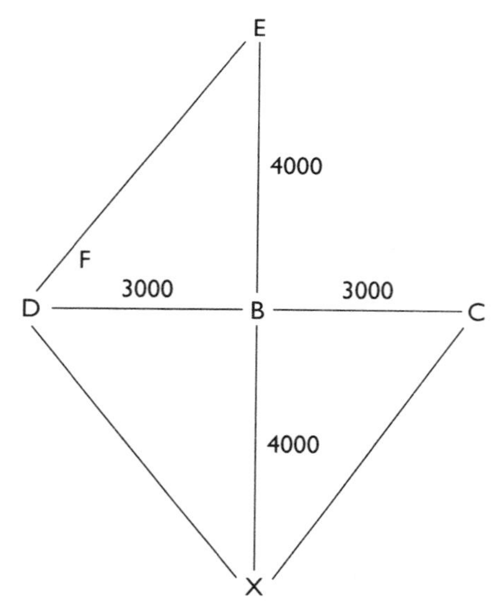

Since CB and BD are collinear, F is to the north of C, and E is closer than X to F, we can pinpoint the Groton.

CBX is also a right triangle. From the facts so far cited, C and D might be the same point. But that is not possible, because we hear that D is closer to some other point (F) than C is. This implies that DB and CB are collinear.

Now, EBD also forms a right triangle. Since E is closer to the *Groton* than X is, EBD and XBD are reflections about segment BD, so E and X are 8000 meters apart. Since E is closer to F than X is, E must be north of the BD segment (since F is north of that segment because of the statement relating F to C). Since E is only 1000 meters from the *Groton* and X is 9000 meters from the *Groton*, the segment from X to the *Groton* must include XE as a subsegment. So the *Groton* is 9000 meters due north of X.

Uneasy Peace

28. Epidemiologists

Ted had Toocies originally. Bob is immune. Here is the reasoning: When Leah met Mary on Tuesday, Mary must have had the disease because she met nobody else afterwards and she had the disease at the end. Therefore, when Leah met Bob later on Tuesday, Leah had the disease. But Bob did not get it. So, Bob is immune and Keith is susceptible. Now if Alice, Ellen, Mary, or Leah had had the disease originally, then Keith would have contracted it. That leaves only Ted as the original carrier.

29. MarsRail

The solution to MarsRail is shown in the figure on page 226. Ecco found it after finding solutions yielding 53, 70, 79, and 81 huts. Can you retrace the reasoning?

Further Thinking: There is no proof that 81 is the best possible. Perhaps you can think of a way to do better.

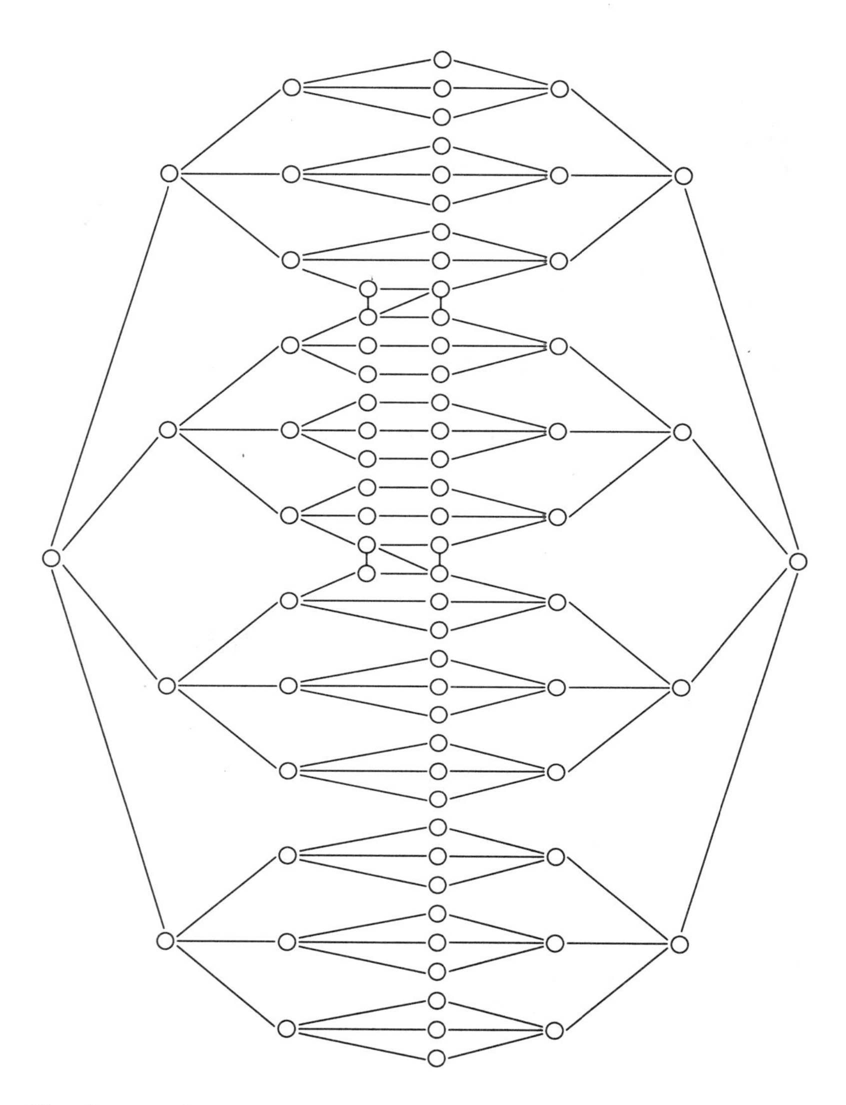

Here is a topology that works.

30. The Prince's Problem

1) No tiling scheme is possible in the 16-by-16-meter room. Ecco proved this by a so-called pigeon-hole argument. Such arguments start from the simple observation that if you must fit many pigeons into few boxes, then some boxes will have several pigeons.

Ecco's reasoning can be cast as a Socratic dialogue. For example, suppose the court mathematician says that he can tile the prince's floor with a fountain in each corner and Ecco says that he cannot.

ECCO: You say that you can do it.

COURT MATHEMATICIAN: Yes, but I won't show you how.

ECCO: It won't change anything if we number each square meter of the prince's floor, will it?

COURT MATHEMATICIAN: Be my guest.

ECCO: I will number each square 0, 1, 2, or 3. I label the top row as follows: 0123012301230123, the second row as 1230123012301230, the third row as 2301230123012301, and so on. That is, each row is a one-digit rotation of the row above [see the figure on p. 228].

COURT MATHEMATICIAN: What does that have to do with my tiling?

ECCO: You notice that there are 64 squares numbered 0, 64 numbered 1, 64 numbered 2, and 64 numbered 3, including the squares with fountains.

COURT MATHEMATICIAN: Yes, so?

ECCO: Any tile you can possibly lay down consumes one 0, one 1, one 2, and one 3.

COURT MATHEMATICIAN: True enough. Any tile that is vertical or horizontal does that.

ECCO: Right. So, your tiles would cover 63 0's, 63 1's, 63 2's, and 63 3's. But you can't do that with one fountain in each corner. As you can see from my drawing, the fountains would cover one 0, one 2, and two 3's (in the upper-right- and lower-left-hand corners). That leaves only 62 3's left.

2) Ecco can show that two diagnonally opposite corners cannot have fountains, so there cannot be a design that uses all 63 tiles and for which more than two corners have fountains in them. The idea is to use a symmetric right-to-left numbering to the left-to-right numbering Ecco used above.

0	1	2	3	0	1	2	3	0	1	2	3	0	1	2	3
1	2	3	0	1	2	3	0	1	2	3	0	1	2	3	0
2	3	0	1	2	3	0	1	2	3	0	1	2	3	0	1
3	0	1	2	3	0	1	2	3	0	1	2	3	0	1	2
0	1	2	3	0	1	2	3	0	1	2	3	0	1	2	3
1	2	3	0	1	2	3	0	1	2	3	0	1	2	3	0
2	3	0	1	2	3	0	1	2	3	0	1	2	3	0	1
3	0	1	2	3	0	1	2	3	0	1	2	3	0	1	2
0	1	2	3	0	1	2	3	0	1	2	3	0	1	2	3
1	2	3	0	1	2	3	0	1	2	3	0	1	2	3	0
2	3	0	1	2	3	0	1	2	3	0	1	2	3	0	1
3	0	1	2	3	0	1	2	3	0	1	2	3	0	1	2
0	1	2	3	0	1	2	3	0	1	2	3	0	1	2	3
1	2	3	0	1	2	3	0	1	2	3	0	1	2	3	0
2	3	0	1	2	3	0	1	2	3	0	1	2	3	0	1
3	0	1	2	3	0	1	2	3	0	1	2	3	0	1	2

The numbering system that Ecco uses to make his argument to the court mathematician. The fountains would have to cover two 3's. Any tile covers one 0, one 1, one 2, and one 3. There are only 62 3's left after laying the fountains, not enough for 63 tiles.

Instead of assigning one number to each square meter, assign two. The first number is assigned as above. For the second number, count as before but right to left. So, the top row could be represented as follows (with first number on top of second):

$$\binom{0}{3}\binom{1}{2}\binom{2}{1}\binom{3}{0}\binom{0}{3}\binom{1}{2}\binom{2}{1}\binom{3}{0}\binom{0}{3}\binom{1}{2}\binom{2}{1}\binom{3}{0}\binom{0}{3}\binom{1}{2}\binom{2}{1}\binom{3}{0}$$

In succeeding rows, both numbers would rotate. So, the second row would have a numbering of:

$$\binom{1}{0}\binom{2}{3}\binom{3}{2}\binom{0}{1}\binom{1}{0}\binom{2}{3}\binom{3}{2}\binom{0}{1}\binom{1}{0}\binom{2}{3}\binom{3}{2}\binom{0}{1}\binom{1}{0}\binom{2}{3}\binom{3}{2}\binom{0}{1}$$

Each tile will cover 0, 1, 2, and 3 of both the first (top) and second (bottom) numbers. So, if the upper-left-hand-corner square meter and lower-right-hand-corner square meter have fountains, then two square meters with bottom number 3 are missing. If the upper-right-hand-corner square meter and lower-left-hand-corner square meter have fountains, then two square meters with top number 3 are missing. There is no way that the 63 tiles can cover the rest of the floor. So, at most one of each diagonally opposite pair can have a fountain. Therefore, at most two corners can have fountains.

In fact, a simple design that works is to put a pair of adjacent fountains in the top left-hand corner and a pair in the top right-hand corner.

3) In a 10-by-10-meter room, there is a simple solution to arranging the tiles and fountains. Create four 5-meter-by-5-meter blocks in each corner. Here is how to do it in the lower left-hand corner: have two tiles pointing away from the corner in both directions, forming an L shape. That leaves a 4-meter-by-4-meter square that can be tiled either horizontally or vertically.

Without the fountains, tiling the 10-by-10-meter room would not be possible. Using the same numbering argument as above, there would be 26 squares with 1's. Twenty-five tiles could never cover them.

31. Oil and Water

Recall that the shortest distance between two points is a straight line. (We can ignore the earth's curvature for these small distances.) Continue the line from the Ace rig past its dock so that the length is doubled. Where the line ends inland is the "reflection" of Ace on land.

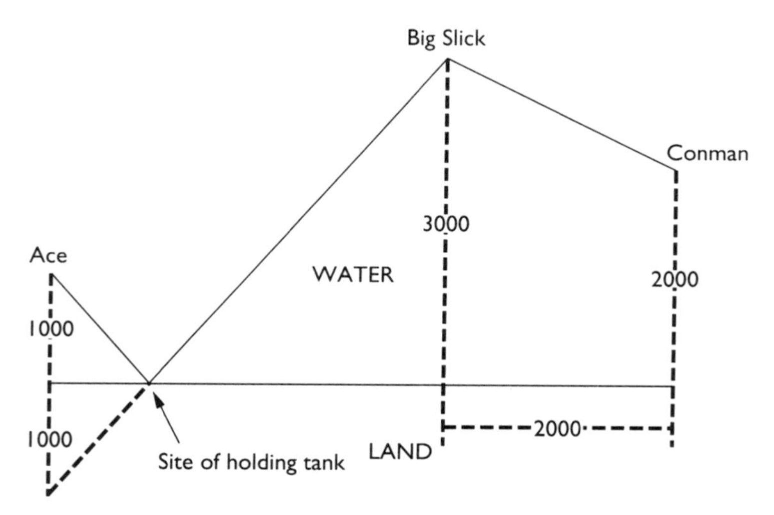

Total length of the pipe between Big Slick and Ace is 5000 meters long, equal to the hyponteneuse of a right triangle with sides of 3000 and 4000 meters. The pipe from Conman to Big Slick is less than 2300 meters long.

Now, draw a straight line from the reflection of Ace to the Big Slick rig. Where the line intersects the land is where the holding tank should go. Using that technique, install the holding tank between Big Slick and Ace and have a pipe from Conman to Big Slick as shown in the figure on this page. This gives a length of $5000 + (5^{1/2} \times 1000)$, or a little under 7250 meters.

Further Thinking: Though the calculation gets messy, you might consider the more economical designs that would result if you were not constrained to join flows at a rig. North and east of Big Slick is one good possible joining point.

32. Pomp But No Power?

1) If the vice president always voted, clearly he would have exactly the same power as a senator, by symmetry. Leaving aside the vice president, all votes end either in a tie or with one side having at least two more votes than the other. In the first case, the vice president

can in fact vote. In the second case, we would allow him to vote—what he did would not matter. So, the vice president has the same power as a senator.

2) When senators abstain, the situation is worse for the vice president, since if an odd number cast nonabstaining votes, the Vice President can never exercise his vote. For example, in the situation that there are 61 nonabstentions and the vote is 30 to 30, the last senator can determine the outcome no matter what the vice president does.

3) Under the rules that Gear proposes, the vice president has more power than a senator. Consider a situation in which there are 50 senators for and 49 senators against a given bill. So, the vice president and one senator remain. No matter what the senator does, the vice president can decide the outcome of the vote. If there are an odd number of senators who vote, the vice president still dominates a senator sometimes. For example, consider the situation in which there are 61 nonabstaining senators and the vote is 30 to 30 with one senator and the vice president still to vote. No matter how the senator votes, the Vice President can determine the outcome of the election. There is no situation in which a senator can force either outcome no matter what the vice-president does.

Further Reading: The inspiration for this puzzle and the next comes from the very nice book *For All Practical Purposes: Introduction to Contemporary Mathematics,* published by the Consortium for Mathematics and Its Applications (New York: W. H. Freeman and Co., 1988).

33. Power Grab

1) No, the proportional scheme leaves Derchev without any power. Derchev can never affect an outcome since any two of Abinev, Brezmev, and Carpow form a winning coalition.

2) No scheme can work. Since Abinev and Brezmev have the same power, neither of them alone can ever prevail. However, Abinev and Brezmev together must win, since they have more power than Carpow and Derchev. If Abinev and Carpow combined also win, then Derchev has no power. If Abinev and Carpow together do not win,

then Carpow has the same power as Derchev, which is also unacceptable.

3) Yes, there is a division that works nicely. Divide the district of Abinev into Abinev the Greater with 3000 voters and Abinev the Lesser with 1000 voters. Similarly, divide Brezmev into Brezmev the Greater with 3000 voters and Brezmev the Lesser with 1000 voters. Give each representative a vote that is directly weighted to the number of voters in his subdistrict. Everyone has power, since Abinev the Greater, Abinev the Lesser, Brezmev the Lesser, and Derchev will win, whereas if any of Abinev the Lesser, Brezmev the Lesser, or Derchev defects, then the other side will win. Moreover, Abinev the Greater and Brezmev the Greater are each more powerful than Carpow, since Abinev the Greater and Brezmev the Greater together will win, but Abinev the Greater and Carpow will not. Finally, Carpow is more powerful than Abinev the Lesser, Brezmev the Lesser, or Derchev, since Abinev the Greater, Carpow, and Abinev the Lesser will win whereas Abinev the Greater, Brezmev the Lesser, and Abinev the Lesser will not.

A Secret Society

34. Odd Voters

In the solutions below, represent yea as 1 and nay as 0.

1) Suppose member B wants to vote 1. B gives a random bit (determined, say, by a coin flip) to each member he speaks to. If the number of 1's sent to the other members is odd, then B gives himself a 0; otherwise, B gives himself a 1. Thus, the total number of 1's that B sends is odd (including the one B sent to himself). Similarly, if B wants to vote 0, then B ensures that the total number of 1's B sends is even (again, including that sent to himself).

In phase 2, B decides 1 if he has received (including from himself) an odd number of 1's. Otherwise, B decides 0.

In phase 3, B votes as he decided in phase 2. Even though B may vote differently from his original choice, the result of the entire election will be the same (odd or even) as it would have been originally. (It may take some thought to see this: notice that B contributes an odd number of 1's whenever he would have voted 1 and an even number of 1's otherwise.)

2) Divide all members into groups of 7, so each member does phase 1 as above, but only to his or her group of 7. (If the total number is not a multiple of 7, then one group may be larger than 7, but no larger than 13.)

Phases 2 and 3 are as before.

3) Suppose member C sends faxes to three other members, giving them random bits as in phase 1 of the first and second solutions. But instead of receiving from those members, C receives from three different ones. Then when C finally votes, her original vote can be determined only by asking the three who sent to her and the three to whom she sent. Breaking this method requires six colluders.

Further Reading: For more discussion of the connection between mathematics and maintaining the privacy of votes, the reader may consult *The Mathematical Tourist: Snapshots of Modern Mathematics* by Ivars Peterson (New York: W. H. Freeman and Co., 1988).

35. Polling the Oddists

1) Suppose the liar decides to lie about one particular Oddist but no others. The only way we can be sure about that Oddist's vote is by designating three pollsters to interview him or her. Since we don't know in advance who the lying pollster is or which Oddist the pollster will lie about, every Oddist must be interviewed by three pollsters. Since each pollster interviews at most 50 Oddists, we conclude that four pollsters are necessary, though it is not yet clear that four are enough.

2) Using Professor Scarlet's assignment of pollsters to Oddists, the pollsters must do better than report back the total number of yeas. Suppose pollster A says there was one yea; pollster B, one; pollster C,

one; and pollster D, two. Two scenarios could explain these results (there are many more, but these two will establish our argument): Oddist 31 alone votes yea and pollster D lies, or Oddists 5 and 55 vote yea and pollster C lies. Since we have no way to choose between these two scenarios, we can only guess whether the number of yeas is even or odd.

3) Keep the professor's assignment of pollsters to Oddists. Encode their results differently.

Consider dividing the 60 Oddists into groups of ten, that is, 1–10, 11–20, and so on up to 51–60. Call these groups I, II, III, IV, V, and VI. Notice that in the professor's scheme each pollster either interviews everyone in a group or nobody in a group. Now for the encoding.

For each group of ten that he or she interviews, each pollster determines whether the number of yeas in that group is even or odd. Call this result the value of that group. If each pollster reports the values of all the groups he or she interviews, then the value of each group would be represented three times in the reports of all the pollsters combined—for example, pollsters B, C, and D would all include a value for group VI.

Each pollster has a value of even or odd from at most five groups. Now the problem is to encode five such values into a number between 0 and 50. Since each value is binary, it can be represented as a 0 (corresponding to "even") and a 1 (corresponding to "odd"). Readers who know about binary representation will recognize immediately that five such values can be represented uniquely into a number between 0 and 31. For readers who are unfamiliar with binary representation, the number for pollster A (the others are similar) will be: (16 × value of group I) + (8 × value of group II) + (4 × value of group III) + (2 × value of group IV) + (1 × value of group V). It is crucial to observe that the resulting number (clearly less than 32, since the value of a group is at most 1) and knowledge of how the number was computed lead one to determine the value for each group.

From the majority view concerning each group, we know the value of that group. Taking the values of all groups, we see whether there are an odd number of 1's. If so, then an odd number of Oddists voted 1.

36. The Hokkaido Post Office Problem

1) Number the participants 1 – 17. Initially, participant 1 (the man who has asked Ecco these questions) sends to everybody else.

After receiving from 1, each in 3 – 17 sends to everyone.

When 2 receives from 1 and from 3 – 17, 2 knows that there are no messages from participant 1 still traveling, so 2 sends to everyone.

2) Once again, participant 1 sends to everyone.

After receiving from 1, each in 4 – 17 sends to everyone. After receiving from 1, participant 3 sends an "all clear" letter to 2 effectively notifying 2 that 3 received the message from 1.

After receiving from 1, 4 – 17, and 3, participant 2 sends to everyone.

After they receive the message from 2, participants 1 and 4 – 17 all send a letter to 3 saying "all clear."

After receiving the message from 2 and the "all clear" messages from 1 and from 4 – 17, participant 3 sends to everyone.

3) Again, participant 1 sends to everyone, starting the protocol.

After receiving the message from 1, participants 2 – 4 send to everyone.

After receiving messages from 1 – 4, each recipient sends an "all clear" message to participants 5 – 8.

After receiving an "all clear" message from everyone, participants 5 – 8 send to everyone.

After receiving messages from 5 – 8, each recipient sends an "all clear" message to participants 9 – 12.

After receiving this "all clear" message from everyone, participants 9 – 12 send to everyone.

After receiving messages from 9 – 12, each recipient sends an "all clear" message to participants 13 – 17.

After receiving this "all clear" message from everyone, participants 13 – 17 send to everyone.

4) Participant 1 sends to everyone through post office A.

After receiving the message from 1, participants 2 – 4 send to everyone through post office A; participants 5 – 8 send to everyone through post office B.

After receiving messages from 1–4 and from 5–8, each recipient sends an "all clear" message to participants 9–17.

After receiving an "all clear" message from everyone, participants 9–12 send through the post office A. After receiving an "all clear" message from everyone, participants 13–17 send through the post office B.

37. Joining the Oddists

1) Here is a sequence of operations that solves the problem:

X ← X xor Y

Y ← X xor Y

X ← X xor Y

To see why this works, notice that no matter what the values of X and Y are, X xor Y xor X = X xor X xor Y = Y xor X xor X = Y. Using this principle, observe what happens when these three operations take place. Let X_i be the initial value of X and Y_i be the initial value of Y. After the first operation, X = X_i xor Y_i. After the second operation, Y = (X_i xor Y_i) xor Y_i = X_i. After the third operation, X = (X_i xor Y_i) xor X_i = Y_i.

2) Here is a solution using 18 such operations. We divide it into three phases.

Phase 1:

Y ← Y xor Z;

X ← X xor Y;

W ← W xor X;

V ← V xor W;

U ← U xor V;

T ← T xor U.

Phase 2:

$Z \leftarrow Y$ xor Z;

$Y \leftarrow X$ xor Y;

$X \leftarrow W$ xor X;

$W \leftarrow V$ xor W;

$V \leftarrow U$ xor V;

$U \leftarrow T$ xor U.

At this point, every storage location except T has its correct value. T has an exclusive OR of all the original values.

Phase 3:

$T \leftarrow T$ xor U;

$T \leftarrow T$ xor V;

$T \leftarrow T$ xor W;

$T \leftarrow T$ xor X;

$T \leftarrow T$ xor Y;

$T \leftarrow T$ xor Z.

This makes 18 operations. Another approach is to use the solution to the first puzzle in a more straightforward way: first swap Y with Z, then X with Y, then W with X, then V with W, then U with V, and finally T with U.

3) If all the assignments can be done in parallel, the solution is simple:

$Z \leftarrow Y$;

$Y \leftarrow X$;

$X \leftarrow W$;

$W \leftarrow V$;

$V \leftarrow U$;

U ← T;

T ← Z.

38. Oddist Summer Training

1) To check whether the number of 1's in the exclusive OR of the two numbers is odd, Akiko sends one postcard to Shu containing a 1 if she has an odd number of 1's and a 0 otherwise. If exactly one of them has an odd number of 1's, then the exclusive OR will too. Otherwise, the exclusive OR will have an even number of 1's. Another odd fact?

2) It may seem necessary to transfer all of Akiko's bits to Shu so Shu can compute the sum. However, we can take advantage of the fact that messages are guaranteed to arrive within one hour. We use the silences to encode information.

Break down the Akiko's 17 bits into five groups of three and one group of two. Akiko will take four hours to send each group of three bits. During at least three of those four hours she will be silent symbolized by #. In the remaining hour she may send a postcard with either a 0 or a 1. Here is one of many encodings:

THREE-BIT VALUE	ENCODING
000	####
001	###0
010	###1
011	##0#
100	##1#
101	#0##
110	#1##
111	0###

Sending the first five groups requires 20 hours using this approach. The last two bits can be sent in two hours using the following encoding:

TWO-BIT VALUE	ENCODING
00	##
01	#0
10	#1
11	0#

3) Akiko and Shu could play a game lasting 2^{16} hours, in which Akiko would send either a 1 or a 0 at hour $n + 1$, if the binary value of the low order 16 bits were n. She would send a 1 if the highest order bit were 1 and a 0 otherwise. The game could last as long as about 7½ years.

Concentric Conspiracies

39. Plea from a Fugitive

Hint: The vowels collapse; all else is easy.

Ecco,

Fate writes strange scripts for us. Today you are the hero and I am the reviled criminal. I am a pirate, kidnapper, and extortionist, they say. It's all true, I suppose.

But above all, I am guilty of stupidity. So are you. You see, Ecco, we still have much in common. We are both pawns in a game of state control. Using me as an excuse, the authorities have begun spontaneous searches and seizures thanks to the new Internal Security Plan I. Internal Security Plan II is ready to go into effect. It will mean total surveillance and universal censorship. The future is clear for those who dare to look.

In order to forestall Internal Security Plan II, I will surrender. They will bring me to trial. All the facts are against me. But there is one small incongruity that may persuade you to act.

U.S. spy ships possess a hidden transmitter with a range of 1000 miles that broadcasts the ship's exact position. The National Security Agency has the capability to decode that broadcast. If you are skeptical, refer to last December's *Defense Electronics,* page 81.

Ecco, if you do decide to act, be prepared for anything.

BB

42. Nightly News

This tape was recorded in a small situation room used by members of the National Security Council. By an old presidential directive, the room is always bugged. Here are some excerpts of conversations that have taken place there. Try to convey this cassette to Paul Savvy at ABC News. He will know what to do with it. Trober

43. Television Treason

The code rotates in the reverse direction of the code that Evangeline and Ecco used in exchanging messages through the *New York Times:*

Nicely done. Believe B to be in serious danger. Escape is hoax. Beware of strangers.

44. Peirce's Beanbag

1) Start by stimulating C, E, and G. After the first year, F and I will also be awake; after two years, D; after three, B. At the end of four years, all regions will be awake.

2) G must be one of the four regions directly stimulated, since G cannot be made awake any other way. Now, consider H and A. If neither is stimulated directly, then both cannot be awake after a year, since no three regions are connected to both. If exactly one of H and